SpringerBriefs in Applied Sciences and Technology

For further volumes:
http://www.springer.com/series/8884

Emily S. C. Ching

Statistics and Scaling in Turbulent Rayleigh-Bénard Convection

 Springer

Emily S. C. Ching
The Chinese University of Hong Kong
Hong Kong
Hong Kong SAR

ISSN 2191-530X   ISSN 2191-5318 (electronic)
ISBN 978-981-4560-22-1 ISBN 978-981-4560-23-8 (eBook)
DOI 10.1007/978-981-4560-23-8
Springer Singapore Heidelberg New York Dordrecht London

Library of Congress Control Number: 2013944533

Printed on acid-free paper

Springer is part of Springer Science+Business Media (www.springer.com)

Preface

When a fluid is heated, its density decreases due to expansion. Hotter fluid, therefore, has a tendency to rise and cooler fluid to fall. The onset of thermal convective motion occurs when the driving force due to the temperature difference is large enough to overcome the dampening effects of fluid viscosity and thermal diffusivity. Above the onset of convection, the velocity of the fluid increases with the temperature difference, and the convective fluid motion becomes turbulent when the velocity is sufficiently fast. Thermal convection is a major mechanism of heat transfer and is often used to provide cooling for equipment in engineering or industrial processes. Thermal convection is also ubiquitous in nature. It plays a role in the structure and dynamics of the Earth's atmosphere and in the earth's mantle, it is an important cause for motion of the tectonic plates. Most thermal convective flows in nature and technological applications are turbulent.

Experimental studies of thermal convection have often been carried out in the setting of Rayleigh-Bénard convection. The Rayleigh-Bénard convection system consists of a closed cell of fluid heated from below and cooled from above. Rayleigh-Bénard convection is a classical problem in fluid mechanics with many issues of interest. It is important for understanding stability of fluid flows and is a good paradigm for studying pattern formation near the onset of convection. In the turbulent regime, both velocity and temperature display complex fluctuations in time. The precise details of the fluctuations vary from flow to flow but the statistical characteristics of the fluctuations are reproducible. A key issue of interest is, therefore, to characterize and make sense of the statistics of the fluctuations. The statistics of the velocity and temperature differences, between measurements taken at two points separated by a distance l, can reveal the structure of turbulence. These structure functions often exhibit a power-law dependence on l, indicating scale invariance. Another issue of interest is to understand this scaling behavior. Amid the fluctuations, flow visualization reveals recurring flow structures that are persistent even after long-time averaging. Prominent examples include a large-scale mean circulating flow that spans the whole cell and thermal plumes, which are the flow structures generated by buoyancy near the top and bottom plates of the convective cell. The third issue of interest is to extract these flow structures from measurements and understand their dynamics and structure. Finally, it is of interest to understand how the heat transfer, which is greatly enhanced by the turbulent fluctuations, depends on the state of fluid flow.

In this monograph, we focus our discussion on the statistical properties and scaling behavior of the velocity and temperature fluctuations in three-dimensional turbulent Rayleigh-Bénard convection. In Chap. 1, the problem of Rayleigh-Bénard convection, under the Oberbeck-Boussinesq approximation, is formulated. The derivation of two exact relations from the equations of motion is presented. In the turbulent regime, the velocity and the temperature fields display irregular fluctuations. In Chap. 2, we introduce the basic statistical tools, including probability density functions (PDF) and conditional statistics, for studying fluctuations in general. As the equations of fluid motion are nonlinear, it is generally not feasible to directly calculate statistical quantities of interest from the equations of motion. This is known as the closure problem in turbulence. Because of this, exact implicit relations between different statistical quantities are useful. We derive two implicit formulae that relate the PDF of fluctuations to two conditional means, and discuss how these formulae have been applied to study the temperature fluctuations in turbulent Rayleigh-Bénard convection. In Chap. 3, we introduce the important concept of energy cascade in turbulent flows and the different theories for the scaling behavior of the velocity and temperature fluctuations. We start with the scaling theory for non-buoyant turbulent flows and then discuss how the presence of buoyancy would affect and modify the scaling behavior. A crossover between the two types of scaling behavior is expected to occur at the length scale above which buoyancy is important. We present and discuss the experimental observations of the scaling behavior and examine the validity of these theoretical expectations in Chap. 4. We end this monograph with a summary and outlook in Chap. 5.

I would like to thank Leo Kadanoff and Albert Libchaber for introducing me to the subject, and acknowledge the fruitful collaborations and discussions with Roberto Benzi, Robert Kraichnan, Stephen B. Pope, Itamar Procaccia, Penger Tong, and Ke-Qing Xia.

Contents

Chapter 1
The Rayleigh-Bénard Convection System

Abstract The Rayleigh-Bénard convection system consists of a closed cell of fluid heated from below and cooled from above. We formulate the problem in this chapter. We first introduce the Oberbeck-Boussinesq approximation and derive the equations of motion under this approximation. The boundary conditions are then specified. The dimensionless parameters describing the state of fluid motion are defined. Furthermore, using the equations of motion, we derive two exact relations relating the heat transfer and the mean energy and thermal dissipation rates.

Keywords Rayleigh-Bénard convection · Oberbeck-Boussinesq approximation · Rayleigh number · Nusselt number · Mean energy and Thermal dissipation rates

1.1 The Oberbeck-Boussinesq Equations

We start with the Navier-Stokes equations for an incompressible flow [1]:

$$\rho \frac{D\vec{U}}{Dt} \equiv \rho \left(\frac{\partial \vec{U}}{\partial t} + \vec{U} \cdot \vec{\nabla}\vec{U} \right) = -\vec{\nabla}p + \eta \nabla^2 \vec{U} \tag{1.1}$$

Here, $\vec{U}(\vec{r}, t)$ is the velocity field, which is a function of the position $\vec{r}$ and time t, $p(\vec{r}, t)$ is the pressure field, ρ and η are the density and the coefficient of viscosity of the fluid. Equation (1.1) is obtained by applying Newton's second law on a small fluid element. $D\vec{U}/Dt$ is the acceleration of the fluid element:

$$\frac{D\vec{U}}{Dt} = \lim_{\Delta t \to 0} \frac{\vec{U}(\vec{r} + \Delta\vec{r}, t + \Delta t) - \vec{U}(\vec{r}, t)}{\Delta t} = \frac{\partial \vec{U}}{\partial t} + \vec{U} \cdot \vec{\nabla}\vec{U} \tag{1.2}$$

It is inherently nonlinear because of the fluid motion. The two terms on the RHS of Eq. (1.2) are the forces per unit volume acting on the fluid element: $-\vec{\nabla}p$ is due to

E. S. C. Ching, *Statistics and Scaling in Turbulent Rayleigh-Bénard Convection*, SpringerBriefs in Applied Sciences and Technology, DOI: 10.1007/978-981-4560-23-8_1, © The Author(s) 2014

the pressure gradient while $\eta \nabla^2 \vec{u}$ is due to viscosity. Conservation of mass leads to the continuity equation

$$\frac{\partial \rho}{\partial t} + \vec{\nabla} \cdot (\rho \vec{U}) = 0 \tag{1.3}$$

which reduces to

$$\vec{\nabla} \cdot \vec{U} = 0 \tag{1.4}$$

for an incompressible fluid with constant ρ. When there are variations in the temperature field $T(\vec{r}, t)$, the density of the fluid changes, thus we need to include the force due to gravity in Eq. (1.1):

$$\rho \left(\frac{\partial \vec{U}}{\partial t} + \vec{U} \cdot \vec{\nabla} \vec{U} \right) = -\vec{\nabla} p + \eta \nabla^2 \vec{U} - \rho g \hat{z} \tag{1.5}$$

where g is the acceleration due to gravity, and $\hat{z}$ is the unit vector along the vertical direction. When the temperature difference is small,

$$\rho = \rho_0 + \delta\rho \approx \rho_0 [1 - \alpha(T - T_0)] \tag{1.6}$$

where α is the coefficient of volume expansion of the fluid, $T_0 = \langle T(\vec{r}, t) \rangle_{V,t}$ is the mean temperature of the system, and ρ_0 is the density at T_0. We use $\langle \cdots \rangle_V$ to denote an average over the whole cell, and $\langle \cdots \rangle_t$ to denote an average over time.

In the Oberbeck-Boussinesq approximation, the change in density due to temperature variation, $\delta\rho$, is taken to be small such that it is neglected in the continuity equation Eq. (1.3) and everywhere in Eq. (1.5) except in the term $\rho g \hat{z}$ in the vertical direction. Therefore, in this approximation, all changes in the fluid properties due to temperature variations are neglected except for the change in density that gives rise to a buoyancy force. In particular, Eq. (1.4) remains valid whereas Eq. (1.5) becomes:

$$\rho_0 \left(\frac{\partial \vec{U}}{\partial t} + \vec{U} \cdot \vec{\nabla} \vec{U} \right) = -\vec{\nabla} p + \eta \nabla^2 \vec{U} - \rho_0 [1 - \alpha(T - T_0)] g \hat{z}$$

$$\Rightarrow \qquad \frac{\partial \vec{U}}{\partial t} + \vec{U} \cdot \vec{\nabla} \vec{U} = -\frac{1}{\rho_0} \vec{\nabla} p^* + \nu \nabla^2 \vec{U} + \alpha g \theta \hat{z} \tag{1.7}$$

where $p^* = p + \rho_0 g z$, $\nu = \eta/\rho_0$ is the kinematic viscosity and $\theta(\vec{r}, t) = T(\vec{r}, t) - T_0$ is the temperature deviation from the mean. The equation of motion for the temperature fluctuation is given by

$$\frac{\partial \theta}{\partial t} + \vec{U} \cdot \vec{\nabla} \theta = \kappa \nabla^2 \theta \tag{1.8}$$

where κ is the thermal diffusivity of the fluid. Equations (1.4, 1.7 and 1.8) are the basic equations for Rayleigh-Bénard convection in the Oberbeck-Boussinesq approximation [1]. They are a set of coupled partial differential equations for the velocity and temperature fields.

1.2 Boundary Conditions

To complete the description of the system, we need to specify the boundary conditions for $\vec{U}$ and θ. For definiteness, we consider a cylindrical cell of height H and radius R and use cylindrical coordinates (r, ϕ, z). The velocity field satisfies the no-slip boundary condition:

$$\vec{U}(r, \phi, z = 0, t) = 0 \, ; \qquad \vec{U}(r, \phi, z = H, t) = 0 \, ; \qquad \vec{U}(r = R, \phi, z, t) = 0$$
(1.9)

For thermally insulating lateral sidewalls, the temperature gradient vanishes at the sidewalls, thus we have

$$\frac{\partial \theta}{\partial r}(r = R, \phi, z, t) = 0$$
(1.10)

The boundary conditions for the temperature field at the top and bottom plates depend on how the thermal driving is enforced. One possible way is to keep the bottom plate at a temperature $T_{bot} = T_0 + \Delta/2$ and the top plate at $T_{top} = T_0 - \Delta/2$, such that there is a fixed temperature difference applied across the cell. In this case, we have

$$\theta(r, \phi, z = 0, t) = \Delta/2$$
(1.11)
$$\theta(r, \phi, z = H, t) = -\Delta/2$$
(1.12)

These boundary conditions are often adopted in theoretical and numerical studies. In experiments, it is, however, more common to put the bottom plate in contact with one or a few heaters that supply a constant heat flux and to connect the top plate to a thermostatic bath circulated by cold water. For such experimental setups, the top plate is approximately fixed at temperature T_{top} so Eq. (1.12) holds but the bottom plate is better approximated as a surface with a constant heat flux. Thus instead of Eq. (1.11), we have:

$$\left\langle \frac{\partial \theta}{\partial z}(r, \phi, z = 0) \right\rangle_t = -\frac{Q}{k}$$
(1.13)

where $Q > 0$ is the constant heat flux supplied to the system and k is the thermal conductivity of the fluid. In this case, the temperature of the bottom plate fluctuates in time and space. We denote the mean temperature of the bottom plate, $\langle T(r, \phi, z = 0, t) \rangle_{z=0,t}$, also by T_{bot} such that

$$\langle \theta(r, \phi, z = 0, t) \rangle_{z=0,t} = \Delta/2$$
(1.14)

and the mean temperature difference across the top and bottom plates is Δ. Here $\langle \cdot \rangle_z$ denotes an average over the horizontal plane at height z of the convection cell.

1.3 Two Dimensionless Parameters: Rayleigh and Prandtl Numbers

Using H as a characteristic length scale, Δ as a characteristic temperature scale, we can construct a characteristic velocity scale: $\sqrt{\alpha g \Delta H}$ and a characteristic time scale $\sqrt{H/(\alpha g \Delta)}$. With these characteristic scales, we can rewrite the basic equations in a dimensionless form:

$$\vec{\nabla} \cdot \vec{U} = 0 \tag{1.15}$$

$$\frac{\partial \vec{U}}{\partial t} + \vec{U} \cdot \vec{\nabla} \vec{U} = -\vec{\nabla} p^* + \sqrt{\frac{\text{Pr}}{\text{Ra}}} \nabla^2 \vec{U} + \theta \hat{z} \tag{1.16}$$

$$\frac{\partial \theta}{\partial t} + \vec{U} \cdot \vec{\nabla} \theta = \frac{1}{\sqrt{\text{PrRa}}} \nabla^2 \theta \tag{1.17}$$

We use the same symbols for the dimensionless quantities to simplify the notations. There are two dimensionless parameters: the Rayleigh (Ra) and the Prandtl (Pr) numbers, which are defined as

$$\text{Ra} = \frac{\alpha g \Delta H^3}{\nu \kappa} , \qquad \text{Pr} = \frac{\nu}{\kappa} \tag{1.18}$$

The parameter Ra measures the relative size of the driving effect due to buoyancy to the dampening effects due to viscosity and thermal diffusivity whereas Pr is a ratio of the diffusivity of momentum due to viscosity to thermal diffusivity. For gases, e.g. air, Pr is of order 1 and does not depend much on temperature. For water, Pr is about 6 at $20\,^{\circ}\text{C}$ and decreases with temperature. For liquid metals like mercury, Pr is of the order of 10^{-2} whereas for Earth's mantle, Pr is around 10^{25}. The aspect ratio Γ is defined as the ratio of the horizontal dimension to the height of the cell. For cylindrical convection cells, $\Gamma = 2R/H$. The state of fluid motion is determined by Ra, Pr and Γ. For a given fluid in a given cell, Γ is fixed and Pr is fixed when the mean temperature of the system is maintained at a fixed value. In this case, the state of fluid motion depends on Ra only. When Ra is small, the fluid does not move and heat is transported by conduction. The onset of convective fluid motion occurs at a certain critical value Ra_c. For an infinite layer of fluid, Ra_c is about 1708. When Ra is slightly larger than Ra_c, steady-state convection occurs and temperature at a fixed position is independent of time. As Ra is increased, temperature fluctuations first becomes chaotic in time with measurements at nearby points remain correlated, then the spatial coherence is lost, and turbulence occurs. The transition to turbulent Rayleigh-Bénard convection occurs at different Ra depending on Pr and also on

Γ [2]. Typical values of Ra in the ocean and in the outer part of the Sun are of the order of 10^{20} and 10^{23} respectively, which are well in the turbulent regime.

1.4 Exact Relations

Using the equations of motion and the boundary conditions, two exact relations can be derived [3]. We show explicitly that the same two relations hold for the constant temperature as well as the constant heat flux boundary conditions at the bottom plate. Take the scalar product of Eq. (1.7) with $\vec{U}$ and the product of Eq. (1.8) with θ, then average over the whole convective cell and time, we get:

$$
\frac{1}{2}\frac{d}{dt}\langle(\vec{U}\cdot\vec{U})\rangle_{V,t} + \frac{1}{2}\langle\vec{U}\cdot\vec{\nabla}(\vec{U}\cdot\vec{U})\rangle_{V,t}
$$
$$
= -\frac{1}{\rho_0}\langle\vec{U}\cdot\vec{\nabla}p^*\rangle_{V,t} + \nu\langle\vec{U}\cdot\nabla^2\vec{U}\rangle_{V,t} + \alpha g\langle U_z\theta\rangle_{V,t} \tag{1.19}
$$

$$
\frac{1}{2}\frac{d\langle\theta^2\rangle_{V,t}}{dt} + \frac{1}{2}\langle\vec{U}\cdot\vec{\nabla}(\theta^2)\rangle_{V,t}
$$
$$
=\kappa\langle\theta\nabla^2\theta\rangle_{V,t} = \kappa\langle\vec{\nabla}\cdot(\theta\vec{\nabla}\theta)\rangle_{V,t} - \kappa\langle|\vec{\nabla}\theta|^2\rangle_{V,t} \tag{1.20}
$$

When the flow is in the stationary state, $d\langle\cdot\rangle_{V,t}/dt$ vanishes. Using the incompressibility condition Eq. (1.4) and the no-slip boundary condition Eq. (1.9), we get

$$
\langle\vec{U}\cdot\vec{\nabla}(\vec{U}\cdot\vec{U})\rangle_V = \langle\vec{\nabla}\cdot[\vec{U}(\vec{U}\cdot\vec{U})]\rangle_V = 0 \tag{1.21}
$$
$$
\langle\vec{U}\cdot\vec{\nabla}p^*\rangle_V = \langle\vec{\nabla}\cdot(\vec{U}p^*)\rangle_V = 0 \tag{1.22}
$$
$$
\langle\vec{U}\cdot\vec{\nabla}(\theta^2)\rangle_V = \langle\vec{\nabla}\cdot(\vec{U}\theta^2)\rangle_V = 0 \tag{1.23}
$$

$$
\langle\vec{U}\cdot\nabla^2\vec{U}\rangle_V = \frac{1}{2}\left\langle\vec{\nabla}\cdot\vec{\nabla}(\vec{U}\cdot\vec{U})\right\rangle_V - \sum_{i,j}\left\langle\left(\frac{\partial U_j}{\partial r_i}\right)^2\right\rangle_V
$$
$$
= -\sum_{i,j}\left\langle\left(\frac{\partial U_j}{\partial r_i}\right)^2\right\rangle_V
$$
$$
= -\frac{1}{2}\sum_{i,j}\left\langle\left(\frac{\partial U_i}{\partial r_j} + \frac{\partial U_j}{\partial r_i}\right)^2\right\rangle_V \tag{1.24}
$$

Thus, Eqs. (1.19, 1.20) become:

$$
\frac{\nu}{2}\sum_{i,j}\left\langle\left(\frac{\partial U_i}{\partial r_j} + \frac{\partial U_j}{\partial r_i}\right)^2\right\rangle_{V,t} = \alpha g\langle U_z\theta\rangle_{V,t} \tag{1.25}
$$

$$
\kappa\langle|\vec{\nabla}\theta|^2\rangle_{V,t} = \kappa\langle\vec{\nabla}\cdot(\theta\vec{\nabla}\theta)\rangle_{V,t} \tag{1.26}
$$

The two terms on the left hand side of Eqs. (1.25, 1.26) are respectively the mean energy and thermal dissipation rates. To see this, we consider the case of decaying turbulence when there is no coupling between velocity and temperature and there is no external forcing, i.e., $\Delta = 0$. In this case, $d\langle \cdot \rangle_{V,t}/dt \neq 0$. Instead,

$$\frac{1}{2}\frac{d}{dt}\langle(\vec{U} \cdot \vec{U})\rangle_{V,t} = -\frac{\nu}{2}\sum_{i,j}\left\langle\left(\frac{\partial U_i}{\partial r_j} + \frac{\partial U_j}{\partial r_i}\right)^2\right\rangle_{V,t} \tag{1.27}$$

$$\frac{1}{2}\frac{d}{dt}\langle\theta^2\rangle_{V,t} = -\kappa\langle|\vec{\nabla}\theta|^2\rangle_{V,t} \tag{1.28}$$

Define

$$\epsilon(\vec{r}, t) = \frac{\nu}{2}\sum_{i,j}\left(\frac{\partial U_i}{\partial r_j} + \frac{\partial U_j}{\partial r_i}\right)^2 \tag{1.29}$$

$$\chi(\vec{r}, t) = \kappa|\vec{\nabla}\theta|^2 \tag{1.30}$$

Thus $\langle\epsilon\rangle_{V,t}$ and $\langle\chi\rangle_{V,t}$ measure the average rate of dissipation of turbulent energy and temperature variance, and are known as the mean energy and thermal dissipation rates respectively. Equations (1.25, 1.26) can be rewritten as:

$$\langle\epsilon\rangle_{V,t} = \alpha g\langle U_z\theta\rangle_{V,t} \tag{1.31}$$

$$\langle\chi\rangle_{V,t} = \kappa\langle\vec{\nabla} \cdot (\theta\vec{\nabla}\theta)\rangle_{V,t} \tag{1.32}$$

Next, we show that the two terms on the right hand side of Eqs. (1.31, 1.32) are related to the heat transfer by the fluid. Define the Nusselt number (Nu) as the heat flux transferred by the fluid normalized by the heat flux in the hypothetical situation that the fluid remains at rest under the same boundary conditions. The heat flux transferred by the fluid across any horizontal plane of the convective cell is given by

$$Q = \langle\rho_0 c U_z\theta - k\frac{\partial\theta}{\partial z}\rangle_{z,t} \tag{1.33}$$

where c is the specific heat capacity of the fluid. In the hypothetical situation that there was no fluid motion, the temperature profile would be linear in z: $\theta = \Delta/2 - (\Delta/H)z$, and the heat flux transferred would be $k\Delta/H$. Thus Nu is defined as

$$\text{Nu} = \frac{Q}{k\Delta/H} \tag{1.34}$$

Since the lateral sidewalls are thermally insulating, Q is independent of z. As a result, we have

$$Q = \frac{1}{H}\int_0^H Q dz = \rho_0 c\langle U_z\theta\rangle_{V,t} + \frac{k\Delta}{H} \tag{1.35}$$

where we have used the boundary conditions Eqs. (1.12, 1.11 or 1.14). Thus

$$\text{Nu} = \frac{\langle U_z \theta \rangle_{V,t}}{\kappa \Delta / H} + 1 \tag{1.36}$$

where the relation $\kappa = k/(\rho_0 c)$ is used. Writing $\langle \vec{\nabla} \cdot (\theta \vec{\nabla} \theta) \rangle_V$ in terms of a surface integral and using Eq. (1.10), we get

$$\langle \vec{\nabla} \cdot (\theta \vec{\nabla} \theta) \rangle_V = \frac{1}{H} \left[\left\langle \theta \frac{\partial \theta}{\partial z} \right\rangle_{z=H} - \left\langle \theta \frac{\partial \theta}{\partial z} \right\rangle_{z=0} \right] \tag{1.37}$$

For the top plate, Eqs. (1.12, 1.33) imply

$$\left\langle \theta \frac{\partial \theta}{\partial z} \right\rangle_{z=H,t} = -\frac{\Delta}{2} \left\langle \frac{\partial \theta}{\partial z} \right\rangle_{z=H,t} = \frac{\Delta Q}{2k} \tag{1.38}$$

For the bottom plate, we get the same result for both the conditions of fixed temperature and constant heat flux:

$$\left\langle \theta \frac{\partial \theta}{\partial z} \right\rangle_{z=0,t} = \begin{cases} \dfrac{\Delta}{2} \left\langle \dfrac{\partial \theta}{\partial z} \right\rangle_{z=0,t} = -\dfrac{\Delta Q}{2k} & \text{fixed temperature} \\[3mm] -\dfrac{Q}{k} \langle \theta \rangle_{z=0,t} = -\dfrac{\Delta Q}{2k} & \text{constant heat flux} \end{cases} \tag{1.39}$$

Here, we have used Eqs. (1.11, 1.33) for the case of fixed temperature condition, and Eqs. (1.13, 1.14) for case of constant heat flux. Thus Eqs. (1.37–1.39) together with Eq. (1.34) give

$$\kappa \langle \vec{\nabla} \cdot (\theta \vec{\nabla} \theta) \rangle_{V,t} = \kappa \frac{\Delta^2}{H^2} \text{Nu} \tag{1.40}$$

Substitute Eqs. (1.36, 1.40) into Eqs. (1.25, 1.26), we finally obtain two exact relations:

$$\langle \epsilon \rangle_{V,t} = \alpha g \kappa \frac{\Delta}{H} (\text{Nu} - 1) = \frac{\nu \kappa^2}{H^4} \text{Ra}(\text{Nu} - 1) \tag{1.41}$$

$$\langle \chi \rangle_{V,t} = \frac{\kappa \Delta^2}{H^2} \text{Nu} \tag{1.42}$$

which relate the mean energy and thermal dissipation rates to Nu and Ra.

References

1. L.D. Landau, E.M. Liftshitz, *Fluid Mechanics* (Pergamon Press, Oxford, London, 1987)
2. P. Manneville, Rayleigh-Bénard Convection: Thirty Years of Experimental, Theoretical, and Modeling Work, In *Dynamics of Spatio-Temporal Cellular Structures: Henri Bénard Centenary Review (Springer Tracts in Modern Physics Vol. 207)*,(Eds.) by I. Mutabazi, J.E. Wesfreid, E. Guyon (Springer, New York, 2006), pp. 41–65
3. E.D. Siggia, High Rayleigh number convection. Annu. Rev. Fluid Mech. **26**, 137–168 (1994)

Chapter 2
Statistical Analysis of Turbulent Fluctuations

Abstract We first introduce the basic statistical tools, including probability density functions (PDF) and conditional statistics, for studying fluctuations in general. Then we discuss the closure problem in turbulence. Because of this closure problem, exact implicit relations between different statistical quantities are useful. We derive two implicit results relating the PDF of fluctuations to two conditional means respectively for stationary and statistically homogeneous fluctuations. Furthermore, we discuss how these implicit PDF formulae have been applied to studying the temperature fluctuations in turbulent Rayleigh–Bénard convection, and the implications of the results obtained.

Keywords Probability density function (PDF) · Conditional probability · Closure problem · Implicit PDF formula · Stationarity · Statistical homogeneity

2.1 Basic Statistical Tools

In turbulent Rayleigh–Bénard convection, as in turbulence in general, experimental measurements of the physical quantities of interest such as velocity and temperature at a certain position display fluctuations as a function of time. The details of the fluctuating measurements in turbulent flows are unpredictable but their statistical properties are reproducible. So one aims to understand the statistical properties and not the details of a turbulent fluid flow. In this section, we introduce the basic statistical tools used in the study of turbulence [1, 2] as well as a general stochastic process.

E. S. C. Ching, *Statistics and Scaling in Turbulent Rayleigh-Bénard Convection*,
SpringerBriefs in Applied Sciences and Technology, DOI: 10.1007/978-981-4560-23-8_2,
© The Author(s) 2014

2.1.1 Cumulative Distribution and Probability Density Function (PDF)

The statistics of any fluctuating quantity are characterized by its probability density function (PDF). Consider any real scalar random variable U, which can have any value in the real space. The cumulative distribution function of U is defined by

$$F(x) = \text{Prob}\{U < x\} \tag{2.1}$$

where Prob$\{\ldots\}$ denotes the probability of the event described inside $\{\ldots\}$. It is clear that $U(x)$ is a non-decreasing function of x. The PDF of U, denoted as $P_U(x)$ or simply $P(x)$ when there is no ambiguity of which variable is being discussed, is defined as the derivative of the cumulative distribution function:

$$P(x) = \frac{dF(x)}{dx} \tag{2.2}$$

$P(x) \geq 0$ and satisfies the normalization condition:

$$\int_{-\infty}^{\infty} P(x)dx = 1 \tag{2.3}$$

The probability of U having a value in a certain interval (x_1, x_2) is

$$\text{Prob}\{x_1 \leq U < x_2\} = F(x_2) - F(x_1) = \int_{x_1}^{x_2} P(x)dx \tag{2.4}$$

For an infinitesimal interval, we have

$$\text{Prob}\{x \leq U < x + dx\} = P(x)dx \tag{2.5}$$

Therefore, the PDF of U is the probability of U per unit 'distance' in the sample space and hence the term probability *density* function.

2.1.2 Moments and Characteristic Function

The quantity

$$\langle U^m \rangle = \int_{-\infty}^{\infty} x^m P(x)dx, \quad m = 1, 2, \ldots \tag{2.6}$$

if exists, is called the mth order moment of the random variable U. Here, $\langle \ldots \rangle$ denotes an ensemble average. If $P(x)$ is an infinitely differentiable function, then the existence of the moments is assured. Most random variables of physical interest can be assumed to have finite moments of the first few orders. The first-order moment of U, with $m = 1$, is called the mean value or mean of U. For any function $Q(U)$ of U, the mean value of Q is given by

$$\langle Q(U) \rangle = \int_{-\infty}^{\infty} Q(U = x) P(x) dx \tag{2.7}$$

The fluctuation of U, denoted as u, is the deviation of U from its mean value, and is defined by

$$u = U - \langle U \rangle \tag{2.8}$$

Denote the PDF of u by $P_u(x)$, then we have $P_u(x) = P_U(x + \langle U \rangle)$. It is convenient to define the central moments, which are the moments of the fluctuation:

$$\langle u^m \rangle = \int_{-\infty}^{\infty} x^m P_u(x) dx, \quad m = 1, 2, \ldots \tag{2.9}$$

The first-order central moment vanishes by definition. The second-order central moment is known as the variance of U and the square root of the variance of U is known as the standard deviation of U. The values of the central moments give information about the shape of the PDF. The standard deviation generally gives information about the width of the distribution. It can be seen from Eq. (2.9) that the contribution to the integral by the large values of $|x|$ increases with m. Thus the high order moments are affected more by the 'tails' of the PDF, the regions of $P(x)$ when $x \rightarrow \pm\infty$. For a variable U whose PDF is symmetric about the mean value: $P_u(x) = P_u(-x)$, the odd central moments would vanish. Thus the degree of asymmetry is often measured by the skewness, which is the third-order central moment normalized by the standard deviation and is denoted by $\mathcal{S}$:

$$\mathcal{S} = \frac{\langle u^3 \rangle}{\langle u^2 \rangle^{3/2}} \tag{2.10}$$

The kurtosis or flatness, denoted as $\mathcal{K}$, is the fourth-order central moment normalized by the standard deviation:

$$\mathcal{K} = \frac{\langle u^4 \rangle}{\langle u^2 \rangle^2} \tag{2.11}$$

A PDF that decreases slowly with $|x|$ and thus having flatter tails will have a larger value of $\mathcal{K}$.

The characteristic function of U is defined as the inverse Fourier transform of the PDF $P(x)$:

$$\phi(k) = \int_{-\infty}^{\infty} e^{ikx} P(x)dx \qquad (2.12)$$

and $P(x)$ is given by the Fourier transform of $\phi(k)$:

$$P(x) = \frac{1}{2\pi} \int_{-\infty}^{\infty} e^{-ikx} \phi(k)dk \qquad (2.13)$$

From Eqs. (2.7) and (2.12), it can be seen that

$$\phi(k) = \langle e^{ikU} \rangle \qquad (2.14)$$

Thus if we write e^{ikU} as an infinite series, we have

$$\phi(k) = \sum_{m=1}^{n-1} \langle U^m \rangle \frac{(ik)^m}{m!} + O(k^n), \qquad (2.15)$$

as long as the moments exist. Moreover, the moments can be obtained by differentiating the characteristic function:

$$\langle U^m \rangle = \frac{d^m \phi(k)}{d(ik)^m} \bigg|_{k=0} \qquad (2.16)$$

Because of this, the characteristic function is also called the moment-generating function.

2.1.3 Some Common Examples of PDFs

1. The uniform distribution:
   If U is uniformly distributed in the interval $[a, b)$, then the PDF of U is

$$P(x) = \begin{cases} \frac{1}{b-a} & a \leq x < b \\ 0 & \text{otherwise} \end{cases} \qquad (2.17)$$

2. The Gaussian or normal distribution:
   If U is normally distributed with mean μ and standard deviation σ, then the PDF of U is

$$P(x) = \frac{1}{\sqrt{2\pi}\sigma} e^{-\frac{(x-\mu)^2}{2\sigma^2}} \tag{2.18}$$

3. The log-normal distribution:
   If U is normally distributed with mean μ and standard deviation σ, then another variable V, defined by $V = e^U$, is, by definition, log-normally distributed. The PDF of V can be obtained from that of U as follows. The cumulative distribution function of V is:

$$F_V(y) = \mathrm{Prob}\{V < y\} = \mathrm{Prob}\{U < \ln y\} = F_U(\ln y) \tag{2.19}$$

The PDF of V is given by

$$P_V(y) = \frac{dF_V(y)}{dy} = \frac{dF_U(\ln y)}{dy} = \frac{dF_U(x)}{dx}\frac{dx}{dy} = \frac{1}{y}P_U(\ln y) \tag{2.20}$$

where $x = \ln y$. Hence using Eq. (2.18),

$$P_V(y) = \frac{1}{\sqrt{2\pi}\sigma y} e^{-\frac{(\ln y - \mu)^2}{2\sigma^2}} \tag{2.21}$$

2.1.4 Joint PDF and Conditional Probability

Our discussions of the cumulative distribution function and PDF of a scalar random variable can be generalized to multiple joint scalar random variables in a straightforward manner. A vector-valued random variable, for example the velocity $\vec{U}$ at a particular location and time in a turbulent flow, can be treated as three joint scalar random variables (U_1, U_2, U_3), where $U_i, i = 1, 2, 3$ are the components of $\vec{U}$. As an illustration, we consider two joint scalar variables in the following discussion, and generalization to more variables is straightforward. The joint cumulative distribution of two random variables U_1 and U_2 is defined as

$$F_{U_1,U_2}(x_1, x_2) = \mathrm{Prob}\{U_1 < x_1 \text{ and } U_2 < x_2\} \tag{2.22}$$

and the joint PDF of U_1 and U_2 is given by

$$P_{U_1,U_2}(x_1, x_2) = \frac{\partial^2 F_{U_1,U_2}(x_1, x_2)}{\partial x_1 \partial x_2} \tag{2.23}$$

Moreover,

$$P_{U_1,U_2}(x_1, x_2)dx_1 dx_2$$
$$= \mathrm{Prob}\{x_1 \leq U_1 < x_1 + dx_1 \text{ and } x_2 \leq U_2 < x_2 + dx_2\} \tag{2.24}$$

The PDF of U_1, denoted by $P_{U_1}(x_1)$, and the PDF of U_2, denoted by $P_{U_2}(x_2)$, can thus be obtained from their joint PDF by

$$P_{U_1}(x_1) = \int_{-\infty}^{\infty} P_{U_1,U_2}(x_1, x_2)dx_2 \tag{2.25}$$

$$P_{U_2}(x_2) = \int_{-\infty}^{\infty} P_{U_1,U_2}(x_1, x_2)dx_1 \tag{2.26}$$

If $Q(U_1, U_2)$ is a function of U_1 and U_2, then its mean is given by

$$\langle Q(U_1, U_2)\rangle = \int_{-\infty}^{\infty} \int_{-\infty}^{\infty} Q(x_1, x_2) P_{U_1,U_2}(x_1, x_2)dx_1dx_2 \tag{2.27}$$

Similarly, the moments of the joint PDF are defined as:

$$\langle U_1^m U_2^n \rangle = \int_{-\infty}^{\infty} \int_{-\infty}^{\infty} x_1^m x_2^n P_{U_1,U_2}(x_1, x_2)dx_1dx_2, \quad m = 1, 2, \ldots \, n = 1, 2, \ldots \tag{2.28}$$

When more than one variable is involved, we can define the conditional probability of some event given that a certain particular event has occurred. That is, out of the original ensemble, one forms a new ensemble in which the particular event has occurred. In particular, the conditional probability density of U_1 given $\{y \leq U_2 < y+dy\}$, denoted as $P_{U_1|U_2}(x|y)$, is the probability density of U_1 in the new ensemble in which $y \leq U_2 < y + dy$. Then

$$P_{U_1|U_2}(x|y)dx = \frac{\text{Prob}\{x \leq U_1 < x + dx \text{ and } y \leq U_2 < y+dy\}}{\text{Prob}\{ y \leq U_2 < y+dy\}} \tag{2.29}$$

and we have

$$P_{U_1|U_2}(x|y) = \frac{P_{U_1,U_2}(x, y)}{P_{U_2}(y)} \tag{2.30}$$

Similarly, denote the conditional probability density of U_2 given $\{y \leq U_1 < y+dy\}$ as $P_{U_2|U_1}(x|y)$, we have

$$P_{U_2|U_1}(x|y) = \frac{P_{U_1,U_2}(y, x)}{P_{U_1}(y)} \tag{2.31}$$

The conditional moments of U_1 given $\{y \leq U_2 < y + dy\}$ are defined by

$$\langle U_1^m \mid U_2 = y \rangle = \int_{-\infty}^{\infty} x^m P_{U_1|U_2}(x|y)dx \quad m = 1, 2, \ldots \tag{2.32}$$

Similarly, the conditional moments of U_2 given $\{y \leq U_1 < y + dy\}$ are defined by

$$\langle U_2^m \mid U_1 = y \rangle = \int_{-\infty}^{\infty} x^m P_{U_2|U_1}(x|y)dx \quad m = 1, 2, \ldots \tag{2.33}$$

If U_1 and U_2 are statistically independent of one another, then $P_{U_1|U_2}(x|y) = P_{U_1}(x)$ and $P_{U_2|U_1}(x|y) = P_{U_2}(x)$. As a result,

$$P_{U_1,U_2}(x_1, x_2) = P_{U_1}(x_1) P_{U_2}(x_2) \tag{2.34}$$

Moreover, the conditional moments reduce to the usual moments:

$$\langle U_1^m \mid U_2 = y \rangle = \langle U_1^m \rangle \tag{2.35}$$

$$\langle U_2^m \mid U_1 = y \rangle = \langle U_2^m \rangle \tag{2.36}$$

2.1.5 Random Functions

A scalar or vector-valued random variable, which depends on one or more space or time coordinates, is called a random function or a random or stochastic process. A physical example is the velocity $\vec{U}(\vec{r}, t)$ as a function of position $\vec{r}$ and time t in turbulent flows. To see how a random function is characterized, we use the simple example of a scalar random variable $U(t)$ that depends on t. At a particular instant $t = t_0$, $U(t_0)$ is a random variable and is characterized by a one-time PDF, $P_1(x, t_0)$. Thus to completely characterize $U(t)$, we need the joint PDF of $U(t_i)$ at all instants of time t_i, $i = 1, 2, \ldots$:

$$P_N(x_1, t_1; x_2, t_2; x_3, t_3; \ldots; x_N, t_N) \quad \text{with } N \to \infty \tag{2.37}$$

2.1.5.1 Statistical Symmetries

When a random function obeys certain statistical symmetries such as stationarity, statistical homogeneity, and statistical isotropy, considerable simplification occurs for its statistics. We shall define these statistical symmetries and discuss the corresponding simplifications.

1. Stationarity:
   A random function $U(t)$ is stationary if its joint PDF at all t is invariant under a shift in time:

$$P_N(x_1, t_1 + h; \ldots; x_N, t_N + h) = P_N(x_1, t_1; \ldots; x_N, t_N) \qquad (2.38)$$

for all N, t_i, $i = 1, 2, \ldots N$, and h. In particular, for $N = 1$ and 2:

$$P_1(x, t + h) = P_1(x, t) \qquad (2.39)$$
$$P_2(x_1, t_1 + h; x_2, t_2 + h) = P_2(x_1, t_1; x_2, t_2) \qquad \forall h \qquad (2.40)$$

Take $h = -t$ in Eq. (2.39), we have

$$P_1(x, t) = P_1(x, 0) = P_1(x) \qquad (2.41)$$

Therefore the one-time PDF and all one-time statistics are independent of time t. Take $h = -t_2$ in Eq. (2.40), we have

$$P_2(x_1, t_1; x_2, t_2) = P_2(x_1, t_1 - t_2; x_2, 0) \qquad (2.42)$$

Thus the two-time PDF and thus all two-time statistics depend on the time difference $t_1 - t_2$ only. As an example, the correlation

$$\langle U(t_1)U(t_2)\rangle = \int\limits_{-\infty}^{\infty} \int\limits_{-\infty}^{\infty} x_1 x_2 P_2(x_1, t_1; x_2, t_2) dx_1 dx_2 \qquad (2.43)$$

depends on $t_1 - t_2$ only.
2. Statistical homogeneity:
   A random function $U(\vec{r})$ is statistically homogeneous if its joint PDF at all $\vec{r}$ is invariant under a shift in position:

$$P_N(x_1, \vec{r}_1 + \vec{l}; \ldots; x_N, \vec{r}_N + \vec{l}) = P_N(x_1, \vec{r}_1; \ldots; x_N, \vec{r}_N) \qquad (2.44)$$

for all N, $\vec{r}_i$, $i = 1, 2, \ldots N$, and $\vec{l}$. Analogous to stationary fluctuations, all one-point statistics at $\vec{r}$ are independent of $\vec{r}$, and all two-point statistics at $\vec{r}_1$ and $\vec{r}_2$, e.g., $\langle U(\vec{r}_1)U(\vec{r}_2)\rangle$ depend on $\vec{r}_1 - \vec{r}_2$ only.
3. Statistical isotropy:
   A random function is statistically isotropic if all its statistics are invariant under rotation.

Consider a general random vector function $\vec{U}(\vec{r}, t)$. Define the correlation function of the ith and jth components of $\vec{U}$ as

$$\Gamma_{ij}(\vec{r}_1, t_1; \vec{r}_2, t_2) = \langle u_i(\vec{r}_1, t_1)u_j(\vec{r}_2, t_2)\rangle \qquad (2.45)$$

where $\vec{u}(\vec{r}, t) = \vec{U}(\vec{r}, t) - \langle \vec{U}(\vec{r}, t) \rangle$ is the fluctuation of $\vec{U}$. If $\vec{U}$ is stationary, then Γ_{ij} depends on $t_1 - t_2$ only. If $\vec{U}$ is statistically homogeneous, then Γ_{ij} depends on $\vec{r}_1 - \vec{r}_2$ only. If $\vec{U}$ is also statistically isotropic, then Γ_{ij} depends on $|\vec{r}_1 - \vec{r}_2|$ only.

The averages that we have discussed so far are ensemble averages. In experiments, statistical properties are usually determined by taking average over time. For stationary processes that decorrelate rapidly enough, time averaging over a sufficiently long period of time gives a good approximation as ensemble averaging. To see this, let $U(t)$ be a stationary process and denote the correlation function of its fluctuation $u = U - \langle U \rangle$, $\langle u(t)u(t') \rangle$, by $R(t - t')$. The time average of U over a time interval T is defined by

$$\langle U(t) \rangle_t \equiv \frac{1}{T} \int_0^T U(t) dt \qquad (2.46)$$

Since U is stationary, $\langle U \rangle$ is independent of time t and $\langle U \rangle = \langle\, \langle U \rangle\, \rangle_t$. Thus

$$\langle [\langle U \rangle_t - \langle U \rangle]^2 \rangle = \langle [\langle\, U - \langle U \rangle\, \rangle_t]^2 \rangle$$

$$= \left\langle \frac{1}{T^2} \int_0^T \int_0^T u(t)u(t') dt dt' \right\rangle$$

$$= \frac{1}{T^2} \int_0^T \int_0^T R(t - t') dt dt'$$

$$= \frac{2}{T^2} \int_0^T dt \int_0^t R(t') dt' \le \frac{2}{T} \int_0^\infty |R(t)| dt \qquad (2.47)$$

where we have interchanged ensemble averaging and integration. Thus if $U(t)$ decorrelates rapidly with time such that $\int_0^\infty |R(t)| dt$ is finite, then $\langle U \rangle_t \to \langle U \rangle$ as $T \to \infty$.

2.1.5.2 Spectral Analysis of Random Functions

For a stationary random function $U(t)$, its frequency spectrum $E(f)$ is defined as the Fourier transform of its correlation function:

$$E(f) = \frac{1}{2\pi} \int_{-\infty}^\infty e^{-ifs} \langle u(t)u(t+s) \rangle ds \qquad (2.48)$$

where $u(t)$ is the fluctuation of $U(t)$. The inverse Fourier transform of Eq. (2.48) gives

$$\Gamma(s) \equiv \langle u(t)u(t+s) \rangle = \int_{-\infty}^{\infty} e^{ifs} E(f) df = 2 \int_0^{\infty} \cos(fs) E(f) df \qquad (2.49)$$

Note that stationarity implies $\langle u(t)u(t+s) \rangle = \langle u(t-s)u(t) \rangle$ and thus $E(f) = E(-f)$. Hence

$$\int_0^{\infty} E(f) df = \frac{1}{2} \langle [u(t)]^2 \rangle \qquad (2.50)$$

and $E(f)df$ is the contribution to $(1/2)\langle [u(t)]^2 \rangle$ due to the Fourier modes in the frequency range between f and $f + df$. When $U(t)$ is a velocity component, then $(1/2)\langle [u(t)]^2 \rangle$ would be the energy (per unit mass) due to the fluctuations of this velocity component and thus $E(f)$ is naturally known as the (one-dimensional) energy frequency spectrum of this velocity component.

For a statistically homogeneous random function $\vec{U}(\vec{r}, t)$, its spectral functions are defined by

$$\Phi_{ij}(\vec{k}, t) = \frac{1}{(2\pi)^3} \int R_{ij}(\vec{r}, t) e^{-i\vec{k}\cdot\vec{r}} d\vec{r} \qquad (2.51)$$

where

$$R_{ij}(\vec{r}, t) = \langle u_i(\vec{x}, t) u_j(\vec{x} + \vec{r}, t) \rangle \qquad (2.52)$$

and $\vec{u} = \vec{U} - \langle \vec{U} \rangle$. Similarly, R_{ij} is obtained from the inverse Fourier transform of Eq. (2.51) and we have

$$\langle \vec{u}(\vec{x}, t) \cdot \vec{u}(\vec{x} + \vec{r}, t) \rangle = \sum_{i=1}^{3} R_{ii}(\vec{r}, t) = \sum_{i=1}^{3} \int_{-\infty}^{\infty} \Phi_{ii}(\vec{k}, t) e^{i\vec{k}\cdot\vec{r}} d\vec{k} \qquad (2.53)$$

If $\vec{U}$ is also statistically isotropic, then $R_{ij}(\vec{r}, t) = R_{ij}(r, t)$ and $\Phi_{ij}(\vec{k}, t) = \Phi_{ij}(k, t)$ where $r = |\vec{r}|$ and $k = |\vec{k}|$. Thus,

$$\int_{-\infty}^{\infty} \Phi_{ii}(\vec{k}, t) e^{i\vec{k}\cdot\vec{r}} d\vec{k} = 4\pi \int_0^{\infty} \Phi_{ii}(k, t) k^2 \left(\frac{\sin kr}{kr} \right) dk \qquad (2.54)$$

and

$$\frac{1}{2} \langle \vec{u}(\vec{x}, t) \cdot \vec{u}(\vec{x} + \vec{r}, t) \rangle = \int_0^{\infty} E(k, t) \left(\frac{\sin kr}{kr} \right) dk \qquad (2.55)$$

where

$$E(k, t) = 2\pi k^2 \sum_{i=1}^{3} \Phi_{ii}(k, t) \tag{2.56}$$

is the spatial spectrum. When $\vec{U}$ is the velocity, $E(k, t)dk$ is the contribution to the turbulent kinetic energy (per unit mass), $(1/2)\langle \vec{u} \cdot \vec{u} \rangle$, from the Fourier modes of wave vector with a magnitude in the range between k and $k + dk$. Thus $E(k, t)$ is known as the energy spatial spectrum.

2.2 The Closure Problem of Turbulence

Using the equations of fluid motion, one can derive equations for the mean velocity and other statistical quantities of interest. However, because the equations of motion are nonlinear, the equation for the mean velocity, a first order moment, will contain terms involving the correlations of two velocity components, which are second-order statistical quantities. Likewise, the equation for the second-order velocity correlations will contain the third-order correlations of three velocity components. As a result, we always have one unknown more than the number of equations, rendering a direct calculation of the statistics of turbulent fluctuations from the equations of motion impossible. This is known as the closure problem of turbulence. An explicit illustration of this problem will be presented using the Navier–Stokes equation Eq. (1.1).

Using the Reynolds decomposition, we write

$$\vec{U}(\vec{r}, t) = \langle \vec{U}(\vec{r}, t) \rangle + \vec{u}(\vec{r}, t) \tag{2.57}$$

By definition, $\langle \vec{u} \rangle = 0$. Using Eq. (1.4), we get

$$\vec{\nabla} \cdot \langle \vec{U} \rangle = 0 \quad \text{and} \quad \vec{\nabla} \cdot \vec{u} = 0 \tag{2.58}$$

We note that the order of taking average and differentiation can be interchanged. Taking the average of Eq. (1.1), we obtain the Reynolds equations:

$$\frac{\partial}{\partial t} \langle \vec{U} \rangle + \langle \vec{U} \cdot \vec{\nabla} \vec{U} \rangle = -\frac{1}{\rho} \vec{\nabla} \langle p \rangle + \frac{\eta}{\rho} \nabla^2 \langle \vec{U} \rangle \tag{2.59}$$

and

$$\langle \vec{U} \cdot \vec{\nabla} \vec{U} \rangle = \langle \vec{U} \rangle \cdot \vec{\nabla} \langle \vec{U} \rangle + \langle \vec{u} \cdot \vec{\nabla} \vec{u} \rangle = \langle \vec{U} \rangle \cdot \vec{\nabla} \langle \vec{U} \rangle + \sum_{i=1}^{3} \frac{\partial}{\partial r_i} \langle u_i \vec{u} \rangle \tag{2.60}$$

As a result, the Reynolds equation for the mean velocity $\langle \vec{U} \rangle$ contains the second-order statistical correlations $\langle u_i u_j \rangle$, which are known as Reynolds stresses. Without additional knowledge about the Reynolds stresses, the mean velocity $\langle \vec{U} \rangle$ cannot be solved from the equations of motion. This is a manifestation of the closure problem.

For practical calculation of the statistical quantities, turbulence modeling is often employed. Turbulence modeling involves finding appropriate models of the Reynolds stress, which are hypotheses of the functional relationship between the Reynolds stresses $\langle u_i u_j \rangle$ and the mean velocity $\langle \vec{U} \rangle$. One example is the turbulent viscosity model [3], in which the Reynolds stresses are expressed in terms of the spatial derivatives of the mean velocity:

$$\langle u_i u_j \rangle = \frac{2}{3} K \delta_{ij} - \nu_T \left(\frac{\partial \langle U_i \rangle}{\partial r_j} + \frac{\partial \langle U_j \rangle}{\partial r_i} \right) \tag{2.61}$$

where $K \equiv (1/2)\langle \vec{u} \cdot \vec{u} \rangle$ and $\nu_T > 0$ is known as the turbulent or eddy viscosity, and can generally be a function of $\vec{r}$ and t. Such turbulence modeling often contains empirical parameters and different turbulent-flow problems will require different turbulence models.

Using Eq. (1.1), one can derive [3, 4] an evolution equation for the one-point, one-time PDF of the velocity at the point $\vec{r}$ and time t, denoted by $P_{\vec{U}}(\vec{V}; \vec{r}, t)$. Define

$$f(\vec{V}; \vec{r}, t) \equiv \delta(\vec{U}(\vec{r}, t) - \vec{V}) \tag{2.62}$$

then $P_{\vec{U}}(\vec{V}; \vec{r}, t)$ is given formally by the the ensemble average of $f(\vec{V}; \vec{r}, t)$:

$$P_{\vec{U}}(\vec{V}; \vec{r}, t) = \langle f(\vec{V}; \vec{r}, t) \rangle \tag{2.63}$$

Moreover, for any well-behaved function $g(\vec{r}, t)$, we have

$$\langle g(\vec{r}, t) f(\vec{V}; \vec{r}, t) \rangle = \langle g(\vec{r}, t) | \vec{U}(\vec{r}, t) = \vec{V} \rangle P_{\vec{U}}(\vec{V}; \vec{r}, t) \tag{2.64}$$

where $\langle g(\vec{r}, t) | \vec{U}(\vec{r}, t) = \vec{V} \rangle$ is the conditional mean of $g(\vec{r}, t)$ given $\{\vec{V} \leq \vec{U} \leq \vec{V} + d\vec{V}\}$. Taking the spatial gradient and time derivative of $f(\vec{V}; \vec{r}, t)$ and using chain rule, we get

$$\frac{\partial f}{\partial t} = -\sum_{j=1}^{3} \frac{\partial f}{\partial V_j} \frac{\partial U_j}{\partial t} = -\sum_{j=1}^{3} \frac{\partial}{\partial V_j} \left(f \frac{\partial U_j}{\partial t} \right) \tag{2.65}$$

$$\frac{\partial f}{\partial r_i} = -\sum_{j=1}^{3} \frac{\partial f}{\partial V_j} \frac{\partial U_j}{\partial r_i} = -\sum_{j=1}^{3} \frac{\partial}{\partial V_j} \left(f \frac{\partial U_j}{\partial r_i} \right) \tag{2.66}$$

Here, we have used the independence of $\vec{U}$ and $\vec{V}$. Furthermore,

$$\vec{U}(\vec{r}, t) \cdot \vec{\nabla} f = \vec{\nabla} \cdot (\vec{U} f) = \vec{\nabla} \cdot (\vec{V} f) = \vec{V} \cdot \vec{\nabla} f \tag{2.67}$$

In Eq. (2.67), the first equality follows from incompressibility, the second equality makes use of the property of the delta-function, and the final equality results because $\vec{V}$ is independent of $\vec{r}$. Hence, using Eqs. (2.65)–(2.67), we have

$$\frac{Df}{Dt} = \frac{\partial f}{\partial t} + \vec{V} \cdot \vec{\nabla} f = -\sum_{j=1}^{3} \frac{\partial}{\partial V_j} \left(f \frac{DU_j}{Dt} \right) \tag{2.68}$$

Taking the ensemble average of Eq. (2.68) and using Eq. (2.64) yields

$$\frac{\partial P_{\vec{U}}}{\partial t} + \vec{V} \cdot \vec{\nabla} P_{\vec{U}} = -\sum_{j=1}^{3} \frac{\partial}{\partial V_j} \left(P_{\vec{U}} \left\langle \frac{DU_j}{Dt} | \vec{U} = \vec{V} \right\rangle \right) \tag{2.69}$$

Finally, using the Navier–Stokes equation Eq. (1.1) to substitute DU_j/Dt, we obtain an evolution equation for $P_{\vec{U}}(\vec{V}; \vec{r}, t)$ in terms of the undetermined conditional mean of the sum of viscous dissipation and the pressure gradient:

$$\frac{\partial P_{\vec{U}}}{\partial t} + \vec{V} \cdot \vec{\nabla} P_{\vec{U}} = -\sum_{j=1}^{3} \frac{\partial}{\partial V_j} \left[P_{\vec{U}} \left(\left\langle \nu \nabla^2 U_j - \frac{1}{\rho} \frac{\partial p}{\partial x_j} | \vec{U} = \vec{V} \right\rangle \right) \right] \tag{2.70}$$

To evaluate the conditional mean, the joint PDF of $\vec{U}$ and $\nu \nabla^2 U_j - (1/\rho)\partial p/\partial x_j$ is needed. Thus the equation for the PDF of $\vec{U}$ depends on the joint PDF of $\vec{U}$ and another variable. Similarly, the equation for the joint PDF of two variables would depend on the joint PDF of three variables. The equation for the joint PDF of N variables introduces a higher-order joint PDF of one additional variable thus again there is always one unknown more than the number of equations. This is another manifestation of the closure problem.

The work of studying turbulence via such an infinite hierarchy of equations for the joint PDFs has been initiated by Lundgren [4], Monin [5], and Novikov [6], and the interested reader is referred to the review [7]. With stochastic modeling for the respective conditional means, a class of turbulent models, known as PDF methods [3], have been developed and particularly applied in turbulent reactive flows [8]. Other work has used data obtained from direct numerical simulations to specify the conditional means with an attempt to achieve a direct link between the observed statistical properties and the basic dynamical features of the systems under consideration [7].

2.3 Exact Implicit PDF Formula

When the systems under consideration obey certain statistical symmetries, the PDF equation for a physical variable of interest can be solved to give an implicit result in terms of the conditional means of the time derivatives or spatial gradients of the variable. Such results have been derived for general stationary [9] or statistically homogeneous fluctuations [10]. Because of the closure problem, these implicit PDF formulae are useful for studying turbulent flows. Furthermore, exact results for the conditional means can be obtained directly from the equations of motion in some cases (see Sect. 2.4).

2.3.1 Stationary Fluctuations

Consider a stationary random process $X(t)$ and denote its one-time PDF as $P_X(x)$. A physical example would be the temperature or a velocity component in a stationary turbulent flow measured at a certain fixed spatial location as a function of time t. Since $X(t)$ is stationary, all its one-time statistics are independent of time. Specifically,

$$\left\langle \frac{d}{dt}\left[h(X)\dot{X}\right]\right\rangle = 0 \tag{2.71}$$

for any well-behaved function $h(X)$ of X. This implies

$$\langle h(X)\ddot{X}\rangle = -\langle h'(X)(\dot{X})^2\rangle, \tag{2.72}$$

where $h'(X) \equiv dh(X)/dX$ and an overdot indicates a time derivative. Write the ensemble averages in Eq. (2.72) in terms of integrals of $P_X(x)$, we get:

$$\int_{-\infty}^{\infty} h(x)\langle \ddot{X} \mid X = x\rangle P_X(x)dx = -\int_{-\infty}^{\infty} h'(x)\langle(\dot{X})^2 \mid X = x\rangle P_X(x)dx \tag{2.73}$$

Here, $\langle \ddot{X} \mid X = x\rangle$ is the conditional mean of the second-order derivative of X given $\{x \leq X < x + dx\}$, and $\langle(\dot{X})^2 \mid X = x\rangle$ is the conditional mean of the square of the derivative of X given $\{x \leq X < x + dx\}$. The two conditional means are thus functions of x. After integrating by parts once the integral on the right hand side and using $P_X(x) \to 0$ as $|x| \to \infty$, we get

$$\int_{-\infty}^{\infty} h(x)\left\{\langle \ddot{X} \mid X = x\rangle P_X(x) - \frac{d}{dx}\left[\langle(\dot{X})^2 \mid X = x\rangle P_X(x)\right]\right\} dx \tag{2.74}$$

Since Eq. (2.74) holds for any well-behaved $h(X)$, the expression inside the curly bracket has to be identically zero. This gives a differential equation for $P_X(x)$:

$$\langle \ddot{X} \mid X = x \rangle P_X(x) = \frac{d}{dx}\left[\langle (\dot{X})^2 \mid X = x \rangle P_X(x) \right] \tag{2.75}$$

Solving Eq. (2.75), we obtain:

$$P_X(x) = \frac{C_1}{\langle \dot{X}^2 \mid X = x \rangle} \exp\left(\int_0^x \frac{\langle \ddot{X} \mid X = x' \rangle}{\langle (\dot{X})^2 \mid x' \rangle} \, dx' \right) \tag{2.76}$$

where C_1 is the normalization constant fixed by $\int_{-\infty}^{\infty} P_X(x)dx = 1$. Equation (2.76), which was first derived in Ref. [9], is an exact formula that expresses $P_X(x)$ implicitly in terms of two undetermined conditional means: $\langle (\dot{X})^2 \mid X = x \rangle$ and $\langle \ddot{X} \mid X = x \rangle$. Its derivation requires only stationarity and differentiability of $X(t)$. With Eq. (2.76), we can calculate $P_X(x)$ if $\langle (\dot{X})^2 \mid X = x \rangle$ and $\langle \ddot{X} \mid X = x \rangle$ are known or we can calculate any one of these three functions if the other two are known.

2.3.2 Statistically Homogeneous Fluctuations

We next consider a statistically homogeneous process $Y(\vec{r})$ whose PDF is $P_Y(y)$. We can think of $Y(\vec{r})$ as the turbulent temperature or velocity fluctuation measured as a function of position **r** at a certain instant of time in a statistically homogeneous fluid flow. Because of homogeneity, we have

$$\langle \nabla \cdot [g(Y)\nabla Y] \rangle = 0 \tag{2.77}$$

for any well-behaved function $g(Y)$ of Y. As a result,

$$\langle g(Y)\nabla^2 Y \rangle = -\langle g'(Y)|\nabla Y|^2 \rangle \tag{2.78}$$

where $g'(Y) \equiv dg(Y)/dY$. Equation (2.78) can be similarly written in terms of integrals of $P_Y(y)$:

$$\int g(y) \left\{ \langle \nabla^2 Y \mid Y = y \rangle P_Y(y) - \frac{d}{dy}\left[\langle |\nabla Y|^2 \mid Y = y \rangle P_Y(y) \right] \right\} dy = 0 \quad (2.79)$$

where $\langle \nabla^2 Y \mid Y = y \rangle$ and $\langle |\nabla Y|^2 \mid Y = y \rangle$ are respectively the conditional mean of the Laplacian of Y and the conditional mean of the square of the gradient of Y given $\{y \le Y < y + dy\}$, and are functions of y. The derivation of an implicit formula for $P_Y(y)$ then parallels that for stationary processes. In particular, the manipulations corresponding to (2.74)–(2.76) lead to

$$\langle \nabla^2 Y \mid Y \rangle P_Y(y) - \frac{d}{dy}\left[\langle |\nabla Y|^2 \mid Y = y \rangle P_Y(y) \right] = 0 \qquad (2.80)$$

and

$$P_Y(y) = \frac{C_2}{\langle |\nabla Y|^2 \mid Y = y \rangle} \exp\left(\int_0^y \frac{\langle \nabla^2 Y \mid Y = y' \rangle}{\langle |\nabla Y|^2 \mid Y = y' \rangle} \, dy' \right). \qquad (2.81)$$

Here C_2 is again the normalization constant determined by $\int_{-\infty}^{\infty} P_Y(y) dy = 1$. Equation (2.81) now expresses $P_Y(y)$ implicitly in terms of $\langle \nabla^2 Y | Y = y \rangle$ and $\langle |\nabla Y|^2 \mid Y = y \rangle$.

In turbulent fluid flows, we are often interested in the statistical properties of not only the velocity or temperature fluctuations measured at one position but also the simultaneous difference of velocity or temperature fluctuations between two positions, i.e., the statistical properties of $\delta Y(\vec{r}_1, \vec{r}_2) \equiv Y(\vec{r}_1) - Y(\vec{r}_2)$. Identities like (2.80) and (2.81) relating the PDF to conditional means of ΔY can be similarly obtained. Instead of (2.77), the starting point is

$$\langle (\nabla_{r_1} + \nabla_{r_2}) \cdot [g(\delta Y)(\nabla_{r_1} + \nabla_{r_2})\delta Y] \rangle = 0. \qquad (2.82)$$

Similar manipulations together with a use of $\nabla_{r_1} \cdot \nabla_{r_2} \delta Y \equiv 0$, yield

$$\langle (\nabla_{r_1}^2 + \nabla_{r_2}^2)\delta Y \mid \delta Y = z \rangle P_{\delta Y}(z) - \frac{d}{dz}\left[\langle |(\nabla_{r_1} + \nabla_{r_2})\delta Y|^2 \mid \delta Y = z \rangle P_{\delta Y}(z) \right] = 0 \tag{2.83}$$

and the PDF of δY, denoted by $P_{\Delta Y}(z)$, is given by

$$P_{\Delta Y}(z) = \frac{C_3}{\langle |(\nabla_{r_1} + \nabla_{r_2})\delta Y|^2 \mid \delta Y = z \rangle} \exp\left(\int_0^z \frac{\langle (\nabla_{r_1}^2 + \nabla_{r_2}^2)\delta Y \mid \delta Y = z' \rangle}{\langle |(\nabla_{r_1} + \nabla_{r_2})\delta Y|^2 \mid \delta Y = z' \rangle} \, dz' \right). \tag{2.84}$$

2.4 Some Exact Results for Conditional Means

With the implicit PDF formulae, the problem of understanding the statistics of any stationary or statistically homogeneous fluctuation is equivalent to understanding the two conditional means of time derivatives or spatial gradients of the fluctuation. In general, the calculation of these conditional means directly from the equation of motion is a very difficult problem. In two special cases of homogeneous turbulent flows, exact results for either one of the two conditional means, $\langle \nabla^2 \theta | \theta = x \rangle$ and $\langle |\nabla \theta|^2 \mid \theta = x \rangle$, of the temperature fluctuation $\theta(\vec{r}, t) = T(\vec{r}, t) - T_0$ can be obtained directly from the equation of motion [11]. The PDF of θ is given by Eq. (2.81) with $Y = \theta$:

$$P_\theta(x) = \frac{C_N}{\langle |\nabla\theta|^2 \mid \theta = x \rangle} \exp\left(\int_0^x \frac{\langle \nabla^2\theta \mid \theta = x' \rangle}{\langle |\nabla\theta|^2 \mid \theta = x' \rangle} \, dx' \right), \tag{2.85}$$

where C_N is the normalization constant. Since only one of the two conditional means can be found, P_θ is still undetermined up to the remaining conditional mean. We are going to discuss these exact results in this section.

2.4.1 Decaying Homogeneous Temperature Fluctuations

Consider an incompressible homogeneous fluid flow without any heat source such that the temperature fluctuations are decaying in time. The governing equations of motion are Eqs. (1.4) and (1.8). Multiplying Eq. (1.8) by $2n\theta^{2n-1}$ and taking an ensemble average, we have

$$\frac{d}{dt}\langle\theta^{2n}\rangle + \langle\vec{\nabla}\cdot(\vec{U}\theta^{2n})\rangle = \kappa\langle\nabla^2(\theta^{2n})\rangle - 2n(2n-1)\kappa\langle\theta^{2n-2}|\nabla\theta|^2\rangle \tag{2.86}$$

where we have used Eq. (1.4). Both $\langle\vec{\nabla}\cdot(\vec{U}\theta^{2n})\rangle$ and $\langle\nabla^2(\theta^{2n})\rangle$ vanish because of homogeneity. As a result, we obtain:

$$\frac{d}{dt}\langle\theta^{2n}\rangle = -2n(2n-1)\kappa\langle\theta^{2n-2}|\nabla\theta|^2\rangle \tag{2.87}$$

The right hand side of (2.87) is negative, expressing the decay of temperature fluctuations. The decay rate for the temperature variance is given by $2\kappa\langle|\nabla\theta|^2\rangle/\langle\theta^2\rangle$, which is the ratio of the scalar dissipation to the variance.

Now suppose the normalized fluctuation, $X \equiv \theta/\langle\theta^2\rangle^{1/2}$, reaches a statistically stationary state although θ itself decays in time [12]. As a result,

$$\frac{d\langle X^{2n}\rangle}{dt} = 0, \tag{2.88}$$

which relates the decay rate of the nth moment, $\langle\theta^{2n}\rangle$, to that of the variance, $\langle\theta^2\rangle$:

$$\frac{1}{\langle\theta^2\rangle^n}\frac{d\langle\theta^{2n}\rangle}{dt} - n\frac{\langle\theta^{2n}\rangle}{\langle\theta^2\rangle^{n+1}}\frac{d\langle\theta^2\rangle}{dt} = 0 \tag{2.89}$$

The approach to stationary X could involve some delicate questions. Suppose that the velocity field spectrum and the initial temperature spectrum have a low-cutoff wavenumber k_c. The advection process will induce a tail $k \ll k_c$ in the temperature spectrum. If the flow region is infinite, this tail will move to ever-lower k. If, instead, the flow region is contained in a periodic box, then there is a smallest wavenumber

k_0 of the order of the inverse of the size of the box. In this case, a stationary cascade of X is expected when the low k-tail has saturated at k_0.

Equations (2.87) and (2.89) yield

$$\langle \theta^{2n-1} \nabla^2 \theta \rangle = -\frac{\langle |\nabla\theta|^2 \rangle}{\langle \theta^2 \rangle} \langle \theta^{2n} \rangle, \tag{2.90}$$

which is true for arbitrary n, and thus implies

$$\langle \nabla^2 \theta \mid \theta = x \rangle = -\frac{\langle |\nabla\theta|^2 \rangle}{\langle \theta^2 \rangle} x \tag{2.91}$$

Equation (2.91) is an exact result for the conditional mean $\langle \nabla^2\theta | \theta = x \rangle$, and holds for the special case of homogeneous temperature fluctuations with no heat source and that the temperature fluctuation decays in such a fashion that its normalized fluctuation is stationary. Then $P_\theta(x)$ is determined by $\langle |\nabla\theta|^2 \,|\theta = x \rangle$ only, and it is given by

$$P_\theta(x) = \frac{C_N}{\langle |\nabla\theta|^2 \mid \theta = x \rangle} \; \exp\left[-\frac{\langle |\nabla\theta|^2 \rangle}{\langle \theta^2 \rangle} \int\limits_0^x \frac{x'dx'}{\langle |\nabla\theta|^2 \mid \theta = x' \rangle} \right]. \tag{2.92}$$

Equation (2.92) was derived by Sinai and Yakhot [12] but they did not emphasize the implied linearity of the conditional mean $\langle \nabla^2\theta \mid \theta = x \rangle$ (Eq. (2.91)). This linearity result is independent of the statistics of the homogeneous velocity field, and is a consequence of the linearity of Eq. (1.8) in θ. The conditional mean is no longer linear if nonlinearity is introduced. Suppose there is an additional nonlinear term $f(\theta)$ on the right hand side of Eq. (1.8). The stationarity of the normalized field now requires

$$\langle [\kappa\nabla^2\theta + f(\theta)]\theta^{2n-1} \rangle = \frac{\langle [\kappa\nabla^2\theta + f(\theta)]\theta \rangle}{\langle \theta^2 \rangle} \langle \theta^{2n} \rangle, \tag{2.93}$$

which implies that $\langle \nabla^2\theta | \theta = x \rangle$ is a nonlinear function of x:

$$\kappa\langle \nabla^2\theta \mid \theta = x \rangle = \left[\frac{\langle f(\theta)\theta \rangle - \kappa\langle |\nabla\theta|^2 \rangle}{\langle \theta^2 \rangle} \right] x - f(x). \tag{2.94}$$

2.4.2 Stationary Homogeneous Temperature Fluctuations with a White-in-time Heat Source

When there is a heat source $s(\vec{r}, t)$, the turbulent temperature fluctuations are stationary, and Eq. (1.8) becomes:

$$\frac{\partial \theta}{\partial t} + \vec{U} \cdot \vec{\nabla} \theta = \kappa \nabla^2 \theta + s \tag{2.95}$$

Consider the heat source s to be a homogeneous white-in-time field that satisfies

$$\langle s(\vec{r}, t) \rangle = 0, \quad \langle s(\vec{r}, t) s(\vec{r}, t') \rangle = 2q \delta(t - t') \tag{2.96}$$

Multiplying both sides of (2.95) by a well-behaved function $g(\theta)$ and taking ensemble averages, we get

$$\frac{d}{dt} \langle G(\theta) \rangle + \langle \vec{U} \cdot \vec{\nabla} G(\theta) \rangle = \kappa \langle g(\theta) \nabla^2 \theta \rangle + \langle s g(\theta) \rangle \tag{2.97}$$

where $G'(\theta) = g(\theta)$ and $'$ denotes a derivative with respect to θ. Stationarity implies $d \langle G(\theta) \rangle / dt = 0$, and $\langle \vec{U} \cdot \vec{\nabla} G(\theta) \rangle = 0$ because of homogeneity and incompressibility as in the previous case. Thus we obtain the steady-state balance equation:

$$\kappa \langle g(\theta) \nabla^2 \theta \rangle + \langle s g(\theta) \rangle = 0 \tag{2.98}$$

To evaluate the term $\langle s g(\theta) \rangle$, we use Eq. (2.95) to invoke the equation of motion of $g(\theta)$:

$$\frac{\partial g(\theta)}{\partial t} = -\vec{U} \cdot \vec{\nabla} g(\theta) + \kappa g'(\theta) \nabla^2 \theta + s g'(\theta) \tag{2.99}$$

Integrating this equation from t' to t with $t > t'$, then multiplying $s(\vec{r}, t)$ and taking ensemble average, we find

$$\langle s(\vec{r}, t) g(\theta(\vec{r}, t)) \rangle = \int_{t'}^{t} \langle s(\vec{r}, t) s(\vec{r}, \tilde{t}) g'(\theta(\vec{r}, \tilde{t})) \rangle \, d\tilde{t} \tag{2.100}$$

In obtaining Eq. (2.100), we have used the fact that $\theta(\vec{r}, \tilde{t})$ is independent of $s(\vec{r}, t)$ for $\tilde{t} < t$ because of causality. Let t' tend to t and use the property of the source s (Eq. (2.96)), we then get

$$\langle s g(\theta) \rangle = q \langle g'(\theta) \rangle \tag{2.101}$$

Using Eqs. (2.98) and (2.101) together with Eq. (2.78) for $Y = \theta$, we find

$$\langle g'(\theta) |\nabla \theta|^2 \rangle = \frac{q}{\kappa} \langle g'(\theta) \rangle \tag{2.102}$$

Equation (2.102) shows that the square of the temperature gradient is uncorrelated with any well-behaved function of the temperature field. Such a feature was also found for the vorticity gradient and the vorticity field in two-dimensional turbulence driven by a white-in-time force [13]. Writing Eq. (2.102) in terms of integrals of P_θ and noting that it is valid for any arbitrary well-behaved function $g'(\theta)$, we finally get

$$\langle |\nabla \theta|^2 \mid \theta = x \rangle = \langle |\nabla \theta|^2 \rangle = \frac{q}{\kappa}. \tag{2.103}$$

Equation (2.103) is an exact result for the case of a homogeneous white-in-time heat source. In this case, the conditional mean $\langle |\nabla \theta|^2 \mid \theta = x \rangle$ is independent of x and is equal to q/κ. Moreover, we have

$$P_\theta(x) = C_N \exp\left[\frac{\kappa}{q} \int_0^x \langle \nabla^2 \theta \mid \theta = x' \rangle dx' \right]. \tag{2.104}$$

Equation (2.104) implies that $\langle \nabla^2 \theta \mid \theta = x \rangle$ is positive when x is negative and negative when x is positive if $P_\theta(x)$ is a decreasing function of $|x|$.

2.5 Stationary Temperature Fluctuations in Rayleigh-Bénard Convection

The Rayleigh–Bénard convection system is not homogeneous throughout the whole cell because of the presence of boundaries. Due to the no-slip boundary condition (Eq. (1.9)), there are viscous (velocity) boundary layers near the top and bottom plates as well as near the sidewall of the cylindrical convection cell. In addition, there are thermal (temperature) boundary layers near the top and bottom plates due to the constant temperature or constant heat flux boundary conditions (Eqs. (1.12) and Eq. (1.11) or Eq. (1.13)). Relatively large mean velocity or temperature gradients exist in the viscous or thermal boundary layers while there are vanishingly small mean velocity and temperature gradients at the center of the cell. Thus the system can be divided into the central region, a core bulk region around the cell center, the boundary layers near the sidewall, and the top and bottom plates, and the remaining regions, and different statistical properties and scaling behavior in the different regions are expected. Temperature fluctuations $T(t)$ are measured by a thermistor as a function of time t at a single location. Let $X(t) \equiv (T - \langle T \rangle)/\sqrt{\langle (T - \langle T \rangle)^2 \rangle}$ be the normalized temperature fluctuation, where $\langle \ldots \rangle_t$ is an average over time.

2.5.1 The Central Region

As the flow is driven by a heat source with a constant heat flux, the temperature fluctuations are stationary and not decaying. Thus the derivation leading to Eq. (2.90) for decaying temperature fluctuations is not valid. Ching [14] assumed that a relation similar to Eq. (2.90) but with the spatial gradient replaced by the time derivative:

$$\langle X^{2n-1} \ddot{X} \rangle = -\langle \dot{X}^2 \rangle \langle X^{2n} \rangle \quad \forall n, \tag{2.105}$$

holds in general for stationary homogeneous turbulent temperature fluctuations. This assumption is not obvious and might be justified by considerations of universality of fluctuations in turbulence.

Using Eq. (2.105), we have

$$\langle \ddot{X} \mid X = x \rangle = -\langle \dot{X}^2 \rangle x \tag{2.106}$$

Substituting Eq. (2.106) into Eq. (2.76) gives:

$$P_X(x) = \frac{C_N}{\langle \dot{X}^2 \mid X = x \rangle} \exp\left[-\int_0^x \frac{\langle \dot{X}^2 \rangle x'}{\langle \dot{X}^2 \mid X = x' \rangle} \, dx' \right] \tag{2.107}$$

Equation (2.107) has first been tested [14] to hold for experimental measurements taken at the center of a convection cell of $H=40$ cm and $D=20$ cm, filled with helium gas at about 5 K [15] for a wide range of Ra from $10^7 - 10^{14}$. The PDF of $X(t)$, $P_X(x)$, and the conditional mean $\langle \dot{X}^2 | X = x \rangle$ are calculated from the experimental data. Both quantities are symmetric in x. A theoretical PDF of X, denoted by $P_{th}(x)$, is computed using the calculated conditional mean and Eq. (2.107). $P_{th}(x)$ agrees very well with $P_X(x)$ for all the data studied. The good agreement has been quantified by calculating

$$\sigma \equiv \frac{\int_{-\infty}^{\infty} \{\log_2[\frac{P_{th}(x)}{P_X(x)}]\}^2 [P_X(x)]^{1/2} dx}{\int\limits_{-\infty}^{\infty} [P_X(x)]^{1/2} dx} \tag{2.108}$$

where the weight factor $[P_X(x)]^{1/2}$ is included to take into account the statistical error in computing $P_{th}(x)$. As can be seen in Table 2.1, $\sigma < 0.5$ for the wide range of Ra studied. The linearity of the conditional mean $\langle \ddot{X}|X = x \rangle$ (Eq. (2.106)) has been directly verified [9] using measurements in helium [15]. A verification using more recent measurements in water [16] is shown in Fig. 2.1. As has been shown in Ref. [14], Eq. (2.107), and thus Eq. (2.106), holds also for the normalized temperature

Table 2.1 Values of σ (see Eq. (2.108)) for the comparison of $P_X(x)$ against $P_{th}(x)$	Ra	σ
	6.9×10^6	0.26
	2.1×10^7	0.30
	6.0×10^8	0.44
	4.0×10^9	0.48
	7.3×10^{10}	0.45
	6.0×10^{11}	0.43
	6.7×10^{12}	0.39
	4.1×10^{13}	0.36
	5.8×10^{14}	0.33

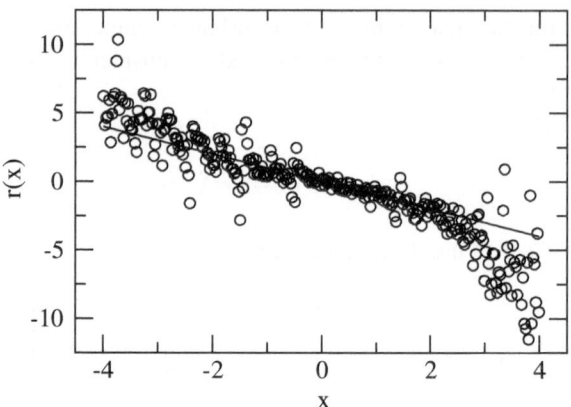

Fig. 2.1 Plot of $r(x) \equiv \langle \ddot{X} \mid X = x \rangle / \langle (\dot{X})^2 \rangle$ as a function of x at the cell center (*circles*) using experimental measurements taken in water at Ra $= 8.3 \times 10^9$ [16]. The *solid line* is $-x$

increments, $X = X_\tau \equiv T_\tau / \sqrt{\langle T_\tau^2 \rangle}$, where $T_\tau(\vec{r}, t) \equiv T(\vec{r}, t + \tau) - T(x, t)$ is the temperature difference separated by a time interval τ, except for very short time separations, as well as for passive temperature data measured in the wake of a slightly heated cylinder. The cylinder was heated so slightly that the buoyancy term was unimportant and temperature acted as a passive scalar. Later work found [17] that Eq. (2.106) holds well also for stationary passive temperature and spanwise vorticity fluctuations obtained in other turbulent shear flows.

An interesting implication of Eq. (2.107) is that the shape of $P_X(x)$, especially its tails at large $|x|$, is determined by the functional dependence of the conditional mean $\langle \dot{X}^2 \mid X = x \rangle$. We shall discuss several cases below.

1. The conditional mean $\langle \dot{X}^2 \mid X = x \rangle$ is independent of x, and thus

$$\langle \dot{X}^2 \mid X = x \rangle = \langle \dot{X}^2 \rangle \tag{2.109}$$

In this case, $P_X(x)$ is a standardized Gaussian with zero mean and unit standard deviation:

$$P_X(x) = \frac{1}{\sqrt{2\pi}} e^{-x^2/2} \tag{2.110}$$

2. The conditional mean $\langle \dot{X}^2 | X = x \rangle$ has a power-law dependence on $|x|$ for large x:

$$\frac{\langle \dot{X}^2 \mid X = x \rangle}{\langle \dot{X}^2 \rangle} \sim \frac{|x|^\alpha}{a} \tag{2.111}$$

for large $|x|$ with $a > 0$ and $0 < \alpha < 2$. Then $P_X(x)$ has stretched exponential tails:

$$P_X(x) \sim e^{-A|x|^{2-\alpha}} \qquad \text{for large } |x| \tag{2.112}$$

where $A = a/(2 - \alpha)$. For the special case of

$$\frac{\langle \dot{X}^2 \mid X = x \rangle}{\langle \dot{X}^2 \rangle} = \frac{1 + B|x|}{1 + B\langle|X|\rangle} \tag{2.113}$$

with $B > 0$, an explicit mathematical form of $P_X(x)$ can be obtained:

$$P_X(x) = C_N e^{-b|x|}(1 + B|x|)^{b/B-1} \tag{2.114}$$

where $b = \langle|X|\rangle + 1/\beta$, and the constants C_N and B are fixed by the normalization condition and $\langle X^2 \rangle = 1$.

3. The conditional mean $\langle \dot{X}^2 \mid X = x \rangle$ is quadratic in x:

$$\frac{\langle \dot{X}^2 \mid X = x \rangle}{\langle \dot{X}^2 \rangle} = \frac{1 + cx^2}{1 + c} \tag{2.115}$$

where $c > 0$. In this case, $P_X(x)$ has algebraic tails [18]:

$$P_X(x) = C_N(1 + cx^2)^{-\left(\frac{1+3c}{2c}\right)} \tag{2.116}$$

and C_N and c are again fixed by the normalization condition and $\langle X^2 \rangle = 1$. PDFs with algebraic tails are related to the observed hyperbolic intermittency in atmospheric dynamics [19].

The PDF of temperature fluctuations at the cell center is Gaussian for Ra below 10^8 in the regime of soft turbulence, and changes to one with exponential tails at higher Ra in the hard-turbulence regime [20]. This change in the PDF can thus be understood as the consequence of the corresponding change in the conditional mean $\langle \dot{X}^2 \mid X = x \rangle$ from Eqs. (2.109) to (2.113). Such a change in the conditional mean has been confirmed directly by the data [10] (see Fig. 2.2). On the other hand, the PDF of the normalized temperature increments, $P_{X_\tau}(x)$, is well approximated by a stretched-exponential $B \exp(-d|x|^\beta)$ with β increasing with τ [21]. This implies that $\langle \dot{X_\tau}^2 \mid X_\tau = x \rangle$ should depend on $|x|$ as $|x|^\alpha$ with $\alpha = 2 - \beta$. This has indeed been found [14].

Using experimental measurements, it has been reported [22, 23] that the velocity and temperature fluctuations in the central region are also statistically homogeneous and isotropic to a good approximation. We note that this can, at best, be an approximation. This is because velocity and temperature fluctuations that are stationary as well as statistically homogeneous are incompatible with the equation of motion for the temperature field (Eq. (1.8)). To show this, recall $\theta(\vec{r}, t) = T(\vec{r}, t) - T_0$, where $T_0 = \langle T(\vec{r}, t) \rangle_{V,t}$ is the average over the whole cell and time. Let $f = \delta(\theta(\vec{r}, t) - w)$. Using results analogous to Eqs. (2.65) and (2.66), we get:

$$\frac{Df}{Dt} = \frac{\partial f}{\partial t} + \vec{\nabla} \cdot (\vec{U} f) = -\frac{\partial}{\partial w}\left[f \frac{D\theta}{Dt}\right] \tag{2.117}$$

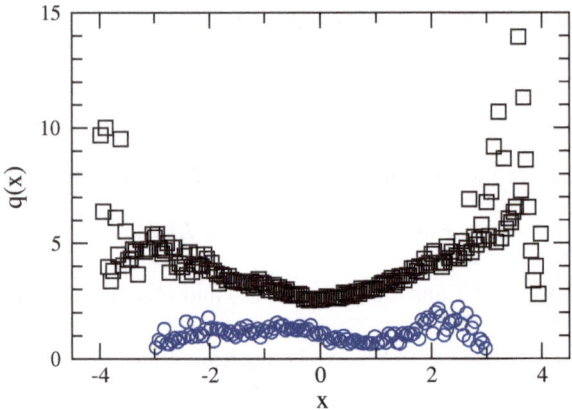

Fig. 2.2 Plot of $q(x) \equiv \langle \dot{X}^2 \mid X = x \rangle / \langle (\dot{X})^2 \rangle$ as a function of x at the cell center (*circles*) using experimental measurements [15] taken in helium at Ra 6.9×10^6 (*circles*) and 7.3×10^{10} (*squares*). The data for Ra=7.3×10^{10} have been shifted up by 2 for clarity

Taking the ensemble average of Eq. (2.117) and using Eq. (1.8) yield an evolution equation for $P_\theta = \langle f \rangle$:

$$\frac{\partial P_\theta}{\partial t} + \vec{\nabla} \cdot [\langle \vec{U} \mid \theta = w \rangle P_\theta] = -\frac{\partial}{\partial w}[\kappa \langle \nabla^2 \theta \mid \theta = w \rangle P_\theta] \qquad (2.118)$$

The temperature fluctuations are stationary thus $\partial P_\theta / \partial t = 0$. If the velocity and temperature fluctuations are also statistically homogeneous, then $\vec{\nabla} \cdot [\langle \vec{U} | \theta = w \rangle P_\theta] = 0$. Thus Eq. (2.118) becomes

$$\langle \nabla^2 \theta \mid \theta = w \rangle P_\theta = C \qquad (2.119)$$

for some constant C. Statistically homogeneity further gives

$$0 = \langle \nabla^2 \theta \rangle = \int\limits_{-\infty}^{\infty} \langle \nabla^2 \theta \mid \theta = w \rangle P_\theta(w) dw \qquad (2.120)$$

which requires that $C = 0$, yielding the unphysical result of a zero $P(\theta)$. Hence, the stationary velocity and temperature fluctuations cannot also be exactly statistically homogeneous in the central region.

2.5.2 Near the Bottom Plate

The temperature fluctuations are stationary but not homogeneous. Thus the PDF of the normalized temperature fluctuations is also a function of position. We assume

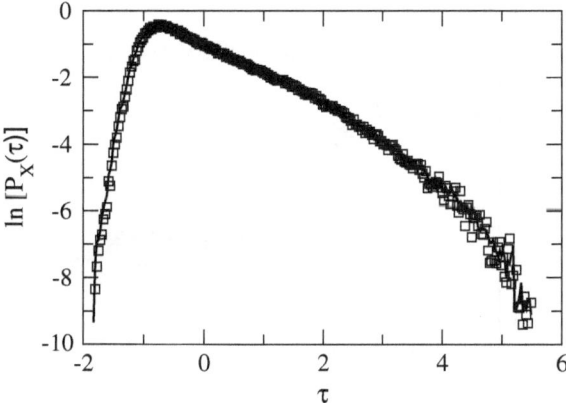

Fig. 2.3 Plot of the PDF of the normalized temperature fluctuations $P_X(\tau)$ (*solid line*) as a function of τ at the center of the bottom plate using experimental measurements taken in water at Ra $= 8.3 \times 10^9$ [16]. $P_{\text{th}}(\tau)$ (squares) is the PDF calculated from Eq. (2.76) using the two conditional means shown in Fig. 2.4. Good agreement can be seen

that the fluctuations are statistically homogeneous on a horizontal plane [24, 25] such that the PDF of the normalized temperature fluctuations is a function of the vertical coordinate z but not the polar coordinates (r, ϕ). In Fig. 2.3, we show $P_X(\tau)$ at the bottom plate. We note that the experimental boundary condition for the bottom plate is given by Eq. (1.13) and not Eq. (1.11) thus there are non-vanishing temperature fluctuations at the bottom plate. It can be seen that unlike the PDF at the cell center, $P_X(\tau)$ at the bottom plate is highly asymmetric in τ. This is caused by much more frequent hotter fluctuations from the bottom plate. Similarly, the two conditional means: $\langle \dot{X}^2 \mid X = \tau \rangle$ and $\langle \ddot{X} \mid X = \tau \rangle$ would generally depend on z too. As shown in Fig. 2.4, $\langle \ddot{X} \mid X = \tau \rangle$ is not linear in τ and $\langle \dot{X}^2 \mid X = \tau \rangle$ is asymmetric. Thus, Eq. (2.105) does not hold for the stationary temperature fluctuations in the thermal boundary layers. Nonetheless, the general result of Eq. (2.76) for stationary fluctuations remains valid, as shown also in Fig. 2.3.

Therefore, in the thermal boundary layers, the functional form of $P_X(\tau)$ is determined by the functional form of both the two conditional means. Moreover, the z-dependence of $P_X(\tau)$ would be reflected by the z-dependence of both $\langle \dot{X}^2 \mid X = \tau \rangle_z$ and $\langle \ddot{X} \mid X = \tau \rangle_z$. Here we use the subscript z to emphasize the z-dependence. On the other hand, in Eq. (2.118), statistical homogeneity on a horizontal plane gives

$$\vec{\nabla} \cdot [\langle \vec{U} \mid \theta = w \rangle_z P_\theta] = \frac{\partial}{\partial z}[\langle U_z \mid \theta = w \rangle_z P_\theta] \qquad (2.121)$$

Thus $P_\theta(w)$ satisfies

$$\frac{\partial}{\partial z}[\langle U_z \mid \theta = w \rangle_z P_\theta] + \frac{\partial}{\partial w}[\kappa \langle \nabla^2 \theta \mid \theta = w \rangle_z P_\theta] = 0 \qquad (2.122)$$

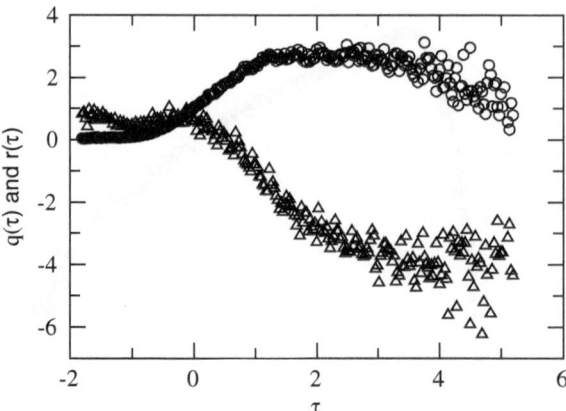

Fig. 2.4 Plot of the conditional means: $q(\tau) = \langle \dot{X}^2 \mid X = \tau \rangle$ (*circles*) and $r(\tau) = \langle \ddot{X} \mid X = \tau \rangle$ (*triangles*) as a function of τ at the center of the bottom plate using experimental measurements taken in water at Ra $= 8.3 \times 10^9$ [16]

This equation has been derived in [25]. The conditional mean $\langle U_z \mid \theta = w \rangle$ has been naturally interpreted as the vertical velocity of the thermal plumes [26]. Hence, the temperature statistics in the thermal boundary layers can be related to the dynamics of the thermal plumes. Now $X(t)$ and $\theta(t)$ are related by

$$\theta(t) = X(t)\langle [T(t) - \langle T \rangle_z]^2 \rangle_z^{1/2} + \langle T \rangle_z - T_0 \,, \tag{2.123}$$

thus

$$P_X(\tau) = P_\theta(w)\langle [T(t) - \langle T \rangle_z]^2 \rangle_z^{1/2} \tag{2.124}$$

where $w = \tau \langle [T(t) - \langle T \rangle_z]^2 \rangle_z^{1/2} + \langle T \rangle_z - T_0$. Furthermore, the conditional means $\langle (\dot{X})^2 \mid X = \tau \rangle_z$ and $\langle \ddot{X} \mid X = \tau \rangle_z$ are related to the conditional means $\langle U_z \mid \theta = w \rangle_z$ and $\langle \nabla^2 \theta \mid \theta = w \rangle_z$. In future work, it is worthwhile to pursue this interesting result to shed light on the physical meanings of the conditional means $\langle (\dot{X})^2 \mid X = \tau \rangle_z$ and $\langle \ddot{X} \mid X = \tau \rangle_z$ (see Chap. 5).

References

1. J.L. Lumley, *Stochastic Tools in Tturbulence* (Dover, New York, 2007)
2. M.T. Landahl, E. Mollo-Christensen, *Turbulence and Random Processes in Fluid Mechanics* (Cambridge University Press, New York, 1992)
3. S.B. Pope, *Turbulent Flows* (Cambridge University Press, Cambridge, 2000)
4. T.S. Lundgren, Distribution functions in the statistical theory of turbulence. Phys. Fluids **10**, 969–975 (1967)
5. A.S. Monin, Equations of turbulent motion. Prikl. Mat. Mekh. **31**, 1057–1067 (1967)

6. E.A. Novikov, Kinetic equations for a vortex field. Sov. Phys. Dokl. **12**, 1006–1008 (1968)
7. R. Friedrich, A. Daitche, O. Kamps et al., The Lundgren-Monin-Novikov hierarchy: Kinetic equations for turbulence. C. R. Physique **13**, 929–953 (2012)
8. S.B. Pope, PDF methods for turbulent reactive flows. Prog. Energy Combust. Sci. **11**, 119–192 (1985)
9. S.B. Pope, E.S.C. Ching, Stationary probability density functions in turbulence. Phys. Fluids A **5**, 1529–1531 (1993)
10. E.S.C. Ching, General formula for stationary or statistically homogeneous probability density functions. Phys. Rev. E **53**, 5899–5903; Erratum. Phys. Rev. E **55**, 4830–4830 (1997)
11. E.S.C. Ching, R. Kraichnan, Exact results for conditional means of a passive scalar in certain statistically homogeneous flows. J. Stat. Phys. **93**, 787–795 (1998)
12. Y.G. Sinai, V. Yakhot, Limiting probability distributions of a passive scalar in a random velocity field. Phys. Rev. Lett. **63**, 1962–1964 (1989)
13. G. Eyink, Exact results on stationary turbulence in 2D: consequences of vorticity conservation. Phys. D **91**, 97–142 (1996)
14. E.S.C. Ching, Probability densities of turbulent temperature fluctuations. Phys. Rev. Lett. **70**, 283–286 (1993)
15. F. Heslot, B. Castaing, A. Libchaber, Transitions to turbulence in helium gas. Phys. Rev. A **36**, 5870–5873 (1987)
16. X. He, P. Tong, Measurements of the thermal dissipation field in turbulent Rayleigh-Benard convection. Phys. Rev. E **79**, 026306 (2009)
17. J. Mi, R.A. Antonia, General relation for stationary probability density functions. Phys. Rev. E **51**, 4466–4468 (1995)
18. E.S.C. Ching, Y.K. Tsang, Passive scalar conditional statistics in a model of random advection. Phys. Fluids **9**, 1353–1361 (1997)
19. D. Schertzer, S. Lovejoy, in *The dimension and intermittency of atmospheric dynamics*, ed. by L.J.S. Bradbury, F. Durst, B.E. Launder, et al. Turbulent shear flows 4 (Springer, Berlin, 1985), pp. 7–33
20. B. Castaing, G. Gunaratne, F. Heslot et al., Scaling of hard thermal turbulence in Rayleigh-Bénard convection. J. Fluid Mech. **204**, 1–30 (1989)
21. E.S.C. Ching, Probabilities for temperature differences in Rayleigh-Bénard convection. Phys. Rev. A **44**, 3622–3629 (1991)
22. Q. Zhou, C. Sun, K.-Q. Xia, Experimental investigation of homogeneity, isotropy, and circulation of the velocity field in buoyancy-driven turbulence. J. Fluid Mech. **598**, 361–372 (2008)
23. R. Ni, S.-D. Huang, K.-Q. Xia, Local energy dissipation rate balances local heat flux in the center of turbulent thermal convection. Phys. Rev. Lett. **107**, 174503 (2011)
24. A.S. Monin, A.M. Yaglom, *Statistical Fluid Mechanics: Mechanics of Turbulence* (MIT Press, Cambridge, 1975)
25. J. Lülff, M. Wilczek, R. Friedrich, Temperature statistics in turbulent Rayleigh-Béard convection. New J. Phy. **13**, 015002 (2011)
26. E.S.C. Ching, H. Guo, X.D. Shang et al., Extraction of plumes in turbulent thermal convection. Phys. Rev. Lett. **93**, 124501 (2004)

Chapter 3
Phenomenology and Scaling Theories

Abstract The statistics of the velocity and temperature differences, between measurements taken at two points separated by a distance l, can reveal the structure of turbulence. These structure functions often exhibit power laws or scaling laws in l. We introduce the important concept of energy cascade in turbulent flows and the different theories for the scaling behavior of the velocity and temperature fluctuations. We start with the scaling theory for non-buoyant turbulent flows and then discuss how the presence of buoyancy would affect and modify the scaling behavior. A crossover between the two types of scaling behavior is expected to occur at a length scale, the Bolgiano length, above which buoyancy is significant. Furthermore, there are corrections to these scaling theories due to the intermittent nature of turbulent fluctuations, and we discuss the idea of refined similarity hypothesis used to account for these corrections.

Keywords Energy cascade · Kolmogorov scaling · Four-fifth law · Obukhov–Corrsin scaling · Bolgiano–Obukhov scaling · Bolgiano length

3.1 Richardson's Energy Cascade

One important concept of fluid turbulence is the energy cascade introduced by Richardson [1]. Turbulent flows are dissipative thus energy input by external forces is required to maintain a turbulent fluid flow. The characteristic scale of this energy input is typically of the size of the system, known as the integral scale, denoted by l_0. On the other hand, the characteristic scale of energy dissipation by viscous effects, known as the dissipative scale and, denoted by l_d, is much smaller than the integral scale. As a result, there must be a transfer of energy from large to small scales. A transfer of energy between scales indicates an interaction between the Fourier modes of velocity of different wave numbers or scales. This is possible because of the non-linear advection term $\vec{U} \cdot \vec{\nabla}\vec{U}$ in the Navier–Stokes equation. Richardson pictured this

E. S. C. Ching, *Statistics and Scaling in Turbulent Rayleigh-Bénard Convection*,
SpringerBriefs in Applied Sciences and Technology, DOI: 10.1007/978-981-4560-23-8_3,
© The Author(s) 2014

energy transfer as a cascade process. Large eddies of the integral scale are produced
by the external forces. They are unstable due to the nonlinearity of the dynamics and
break up into eddies of smaller scale. These smaller eddies are themselves unstable
and break up into eddies of even smaller scale. This process continues until dissipative effects due to viscosity are significant, and the turbulent kinetic energy is
then dissipated into heat. This picture of energy cascade is succinctly summarized
in Richardson's famous rhyme [1]:

> Big whorls have little whorls
> Which feed on their velocity
> And little whorls have lesser whorls
> And so on to viscosity
> (in the molecular sense)

An inherent feature of the cascade picture is that the energy transfer among scales
is local, that is, the effective energy exchange between modes of different wave
numbers decreases as the ratio of the wave numbers increases. It is expected that
the statistics at the integral scale are determined by the mechanism of energy input
and would vary from flow to flow. Because of the locality of the energy transfer, the
statistics at small scales, scales further down the cascade and far from the integral
scale, are not directly influenced by the mechanism of energy input. Thus this locality
feature of the energy cascade allows for the possibility of universal characteristics
for the statistics of small scales.

3.2 The Kolmogorov 1941 Theory

Based on Richardson's energy cascade, Kolmorogov developed in 1941 a phenom-
enological theory (K41) [2] for the statistics of velocity difference,

$$\delta \vec{U}(\vec{r}, \vec{l}) \equiv \vec{U}(\vec{r} + \vec{l}, t) - \vec{U}(\vec{r}, t) \tag{3.1}$$

measured at the same time and at two positions separated by a displacement vector $\vec{l}$.
There are several hypotheses in the K41 theory. We shall focus on two of them. The
first one is an assumption of statistical homogeneity and isotropy of the small-scale
turbulent motion when, the Reynolds number (Re) is sufficiently high and far from
the boundaries. For statistically homogeneous fluctuations, $\delta U(\vec{r}, \vec{l}) = \delta u(\vec{r}, \vec{l})$ as
$\langle \vec{U}(\vec{r}, t) \rangle = \langle \vec{U}(\vec{r} + \vec{l}, t) \rangle$. Thus under this hypothesis, the statistics of $\delta \vec{u}(\vec{r}, \vec{l})$ do
not depend on $\vec{r}$ nor the direction of $\vec{l}$ but depend only on $l = |\vec{l}|$ for $l \ll l_0$. The
second assumption is that under the same conditions stated in the first assumption,
there exists a range of intermediate length scales in which the statistics of $\delta \vec{u}(l)$ are
uniquely and universally determined by the mean energy transfer rate and l. This
range of intermediate length scales, $l_d \ll l \ll l_0$, is known as the inertial range. The
locality of energy cascade makes it possible that the statistics in the inertial range to
be universal. The mean energy transfer rate is equal to the mean energy dissipation

rate as well as to the mean energy input rate. The mean energy dissipation rate is given by $\langle \epsilon \rangle$, where $\epsilon(\vec{r}, t)$ is defined by Eq. (1.29), and the ensemble average, can be taken as the spatial average in statistically homogeneous turbulent flows.

Using these two hypotheses and dimensional analysis, one therefore obtains

$$\langle \delta \vec{u}(l) \cdot \delta \vec{u}(l) \rangle = C \langle \epsilon \rangle^{2/3} l^{2/3} \tag{3.2}$$

where C is a universal constant. Since

$$\langle \delta \vec{u}(l) \cdot \delta \vec{u}(l) \rangle = 2[\langle \vec{u} \cdot \vec{u} \rangle - \langle \vec{u}(\vec{r} + \vec{l}, t) \cdot \vec{u}(\vec{r}, t) \rangle] \tag{3.3}$$

we obtain

$$\langle \delta \vec{u}(l) \cdot \delta \vec{u}(l) \rangle = 4 \int_0^{\infty} E(k) \left[1 - \left(\frac{\sin kl}{kl} \right) \right] dk \tag{3.4}$$

using Eq. (2.55). As a result, Eq. (3.2) is equivalent to the result that the spatial energy spectrum $E(k)$ follows a $k^{-5/3}$ law over a suitable range of wave number k. There is good experimental support for the $-5/3$ power-law in the energy frequency spectrum $E(f)$ (see [5] for details). Similar arguments can be applied to give higher-order statistics of $\delta \vec{u}(l)$. The longitudinal velocity difference along the direction of the separation, denoted by $\delta u_{\parallel}(l)$, is given by

$$\delta u_{\parallel}(l) \equiv \delta \vec{u}(l) \cdot \frac{\vec{l}}{l} \tag{3.5}$$

Then we have

$$S_p(l) \equiv \langle [\delta u_{\parallel}(l)]^p \rangle = C_p \langle \epsilon \rangle^{p/3} l^{p/3} \tag{3.6}$$

for arbitrary $p > 0$. Here, $S_p(l)$ is known as the longitudinal velocity structure function of order p. A power-law dependence of $S_p(l)$ on l indicates that the inertial-range turbulent statistics are scale-invariant and Eq. (3.6) is a statement of the K41 scaling. For $p = 3$, an exact result can be derived from the Navier–Stokes equation for statistically homogeneous and isotropic flows:

$$S_3(l) = -\frac{4}{5} \langle \epsilon \rangle l \tag{3.7}$$

This exact result is known as the four-fifth law and was derived by Kolmogorov also in 1941 [3]. It is one of the very few exact results for turbulent flows, and will be discussed in details in the next section.

In the above discussion, the K41 scaling is obtained by dimensional analysis based on the requirement that in the inertial range $S_p(l)$ depends only on $\langle \epsilon \rangle$ and the scale l. Thus we have the same scaling behavior for the velocity structure function using $\delta u(l) = |\delta \vec{u}(l)|$:

$$S_p^u(l) \equiv \langle [\delta u(l)]^p \rangle \sim \langle \epsilon \rangle^{p/3} l^{p/3} \tag{3.8}$$

The K41 scaling can also be obtained by the requirement that the rate of energy transfer (per unit mass) is scale-independent in the inertial range. To see this, think of $\delta u(l)$ as the velocity of a turbulent eddy of scale l. The energy transfer rate (per unit mass) at scale l can be estimated as $[\delta u(l)]^2 / t_l$. Here, t_l, known as the eddy turnover time, is the typical time for the eddy of size l to deform or change in energy, and can be estimated as $t_l = l / \delta u(l)$. Requiring the rate of energy transfer to be independent of scale l implies $[\delta u(l)]^3 / l \sim const$ but the mean energy transfer rate has to be equal to the mean energy dissipation rate thus $const = \langle \epsilon \rangle$, hence

$$\frac{[\delta u(l)]^3}{l} \sim \langle \epsilon \rangle \tag{3.9}$$

This gives

$$\delta u(l) \sim \langle \epsilon \rangle^{1/3} l^{1/3} \tag{3.10}$$

which further implies Eq. (3.8).

The dissipative scale l_d can be estimated as the scale at which the rate of dissipation due to viscosity is comparable to $\langle \epsilon \rangle$:

$$\nu \left[\frac{\delta u(l_d)}{l_d} \right]^2 \sim \langle \epsilon \rangle \tag{3.11}$$

Then take l_d to be at the edge of the inertial range such that Eq. (3.10) holds at $l = l_d$. Eliminating $\delta u(l_d)$ from Eqs. (3.10) and (3.11) gives $l_d \sim (\nu^3 / \langle \epsilon \rangle)^{1/4}$. The Kolmogorov dissipative scale η_K is defined as:

$$\eta_K \equiv \left(\frac{\nu^3}{\langle \epsilon \rangle} \right)^{1/4} \tag{3.12}$$

and typically l_d is of the order of $10\eta_K$.

Denote the normalized velocity difference by Y_l:

$$Y_l \equiv \frac{\delta u(l)}{\langle [\delta u(l)]^2 \rangle^{1/2}} \tag{3.13}$$

An important consequence of Eq. (3.8) is that all the moments of Y_l are independent of l. This l-independence follows directly from the proportionality of the scaling exponents $p/3$ of $S_p^u(l)$ to p, and further implies that the PDF of Y_l is independent of l. Thus the K41 theory predicts that the statistics of inertial-range turbulent velocity fluctuations are scale-independent and thus self-similar. Experiments confirmed the power-law dependence or scaling of $S_p^u(l)$ but show that the scaling exponents, defined by $S_p^u(l) \sim l^{\zeta(p)}$, depend on p in a nonlinear fashion. This deviation of the scaling behavior from the K41 prediction is known as anomalous scaling. Furthermore, the discrepancy between the observed scaling exponents $\zeta(p)$ and the predicted values of $p/3$ is known as intermittency corrections as the origin of the correction is believed to be due to the intermittent nature of turbulent fluctuations. The problem of anomalous scaling is a longstanding problem of turbulence and is remained to be solved. We shall discuss one particular idea, the refined similarity hypothesis, proposed by Kolmogorov and Obukhov in Sect. 3.7.

3.3 The Four-Fifth Law

We show in detail the derivation of the exact four-fifth law. We shall follow the treatment in [4] and [5]. In this subsection, we denote $\partial/\partial t$ and $\partial/\partial r_i$ by ∂_t and ∂_i, and adopt the Einstein notation of summation over repeated indices. We write the Navier–Stokes equation Eq. (1.1) with an external force in component form:

$$\partial_t U_i + U_k \partial_k U_i = -\frac{1}{\rho}\partial_i p + \nu \partial_k^2 U_i + f_i \tag{3.14}$$

Here, $\rho \vec{f}$ is the external force per unit volume. Denote quantities evaluated at $\vec{r}\,' = \vec{r} + \vec{l}$ by the same notations with a prime, e.g., $U_i' \equiv U_i(\vec{r}\,', t)$, and $\partial/\partial r_i'$ by ∂_i'. Then taking the ensemble average of the product of U_j' with Eq. (3.14) and U_i with Eq. (3.14) for U_j', we have

$$\partial_t \langle U_i U_j' \rangle = -\partial_k \langle U_k U_i U_j' \rangle - \partial_k' \langle U_k' U_j' U_i \rangle - \frac{1}{\rho}\partial_i \langle U_j' p \rangle - \frac{1}{\rho}\partial_j' \langle U_i p' \rangle$$
$$+ \nu(\partial_k^2 + \partial_k'^2)\langle U_i U_j' \rangle + \langle U_j' f_i \rangle + \langle U_i f_j' \rangle \tag{3.15}$$

Here, we have used the interchangeability of taking derivative and ensemble average, incompressibility, and the derivative of primed quantities with respect to unprimed coordinates vanishes.

Consider turbulent flows that are statistically homogeneous and isotropic. Because of homogeneity and isotropy, the averages of the product of primed and unprimed quantities depend only on $l = |\vec{r}\,' - \vec{r}|$. Therefore,

$$\partial_i \langle \cdot \rangle = -\partial_i' \langle \cdot \rangle = -\partial_{l_i} \langle \cdot \rangle \tag{3.16}$$

Thus

$$\langle U_j' p \rangle = g(l) n_j \tag{3.17}$$

for some function $g(l)$ and $n_j \equiv l_j/l$. Incompressibility implies $0 = \partial_j' \langle U_j' p \rangle = \partial_{l_j} \langle U_j' p \rangle$, and using

$$\partial_{l_i} = \frac{\partial l}{\partial l_i}\partial_l = n_i\partial_l, \quad \partial_{l_i} = \frac{\partial n_k}{\partial l_i}\partial_{n_k} = \frac{1}{l}(\delta_{ik} - n_i n_k)\partial_{n_k} \tag{3.18}$$

we get

$$\frac{dg(l)}{dl} + \frac{2}{l}g(l) = 0 \Rightarrow g(l) = \frac{const}{l^2} \tag{3.19}$$

But $g(l)$ has to be finite at $l = 0$ thus $const = 0$ giving $g(l) = 0$ or $\langle U_i' p \rangle = 0$. Similarly, $\langle U_i p' \rangle = 0$. Define the velocity correlation and structure functions as follows.

$$b_{i,j} = \langle U_i U_j' \rangle \tag{3.20}$$

$$B_{ij} = \langle (U_i' - U_i)(U_j' - U_j) \rangle \tag{3.21}$$

$$b_{ij,m} = \langle U_i U_j U_m' \rangle \tag{3.22}$$

$$B_{ijm} = \langle (U_i' - U_i)(U_j' - U_j)(U_m' - U_m) \rangle \tag{3.23}$$

Statistical homogeneity and isotropy imply that these functions depend only on l. Moreover, $\langle U_i(\vec{r}, t)U_j(\vec{r} + \vec{l}, t) \rangle = \langle U_i(\vec{r} - \vec{l}, t)U_j(\vec{r}, t) \rangle = \langle U_i(\vec{r} + \vec{l}, t)U_j(\vec{r}, t) \rangle$, thus $b_{i,j} = b_{j,i}$ is symmetric in the indices i and j. Therefore, the most general forms for $b_{i,j}$ and $b_{ij,m}$ are:

$$b_{i,j} = A(l)\delta_{ij} + B(l)n_i n_j \tag{3.24}$$

$$b_{ij,m} = C(l)\delta_{ij}n_m + D(l)\left(\delta_{im}n_j + \delta_{jm}n_i\right) + F(l)n_i n_j n_m \tag{3.25}$$

The form in Eq. (3.25) takes into account the symmetry in the indices i and j. Homogeneity implies $\langle U_j' U_k' U_i \rangle = \langle U_j(\vec{r}, t)U_k(\vec{r}, t)U_i(\vec{r} - \vec{l}, t)$, which is equal to $-b_{jk,i}$ using Eq. (3.25). We write Eq. (3.15) for statistically homogeneous and isotropic turbulent flows:

$$\partial_t b_{i,j} = -\partial_k(b_{ki,j} + b_{kj,i}) + 2\nu\partial_k^2 b_{i,j} + \langle U_j' f_i \rangle + \langle U_i f_j' \rangle \tag{3.26}$$

Using Eq. (3.26), a relation between the second- and third-order longitudinal velocity structure functions can be derived and from this relation the four-fifth law follows.

As we are interested in the longitudinal structure functions, we let the x-axis be along the direction of $\vec{l}$. Then we take $i = j = 1$ in Eq. (3.26) and obtain

$$\partial_t b_{\|,\|} = -2\partial_k b_{k\|,\|} + 2\nu\partial_k^2 b_{\|,\|} + \frac{2}{3}\langle \vec{f} \cdot \vec{U} \rangle - \langle \delta f_\| \delta u_\| \rangle \tag{3.27}$$

where $\delta f_i = f_i' - f_i$. The subscript $\|$ denotes the component along the longitudinal direction along $\vec{l}$ and there is no summation over this direction. Next, we relate $\partial_k^2 b_{\|,\|}$ to $S_2(l)$. Now $S_2(l)$ can be written as

$$S_2(l) = \langle \delta u_\parallel^2 \rangle = B_{ij} n_i n_j = \frac{2}{3} \langle \vec{U} \cdot \vec{U} \rangle - 2b_{\parallel,\parallel} = \frac{2}{3} \langle \vec{U} \cdot \vec{U} \rangle - 2(A + B) \quad (3.28)$$

where we have used

$$\langle U_i' U_j' \rangle = \langle U_i U_j \rangle = \frac{1}{3} \langle \vec{U} \cdot \vec{U} \rangle \delta_{ij} \quad (3.29)$$

Using Eqs. (3.16) and (3.18), we get

$$\partial_k^2 b_{i,j} = \left(\frac{d^2 A}{dl^2} + \frac{2}{l} \frac{dA}{dl} + \frac{2}{l^2} B \right) \delta_{ij} + \left(\frac{d^2 B}{dl^2} + \frac{2}{l} \frac{dB}{dl} - \frac{6}{l^2} B \right) n_i n_j \quad (3.30)$$

The incompressibility condition gives $0 = \partial_j' b_{i,j}$, which implies

$$\frac{l}{2} \frac{d}{dl}(A + B) + B = 0 \quad (3.31)$$

Thus

$$\frac{dS_2}{dl} = -2 \frac{d}{dl}(A + B) = \frac{4}{l} B(l) \quad (3.32)$$

and

$$\partial_k^2 b_{i,j} = -\left(\frac{d^2 B}{dl^2} + \frac{4}{l} \frac{dB}{dl} \right) \delta_{ij} + \left(\frac{d^2 B}{dl^2} + \frac{2}{l} \frac{dB}{dl} - \frac{6}{l^2} B \right) n_i n_j$$

$$\Rightarrow \quad \partial_k^2 b_{\parallel,\parallel} = -\frac{2}{l^4} \frac{d(l^3 B)}{dl} = -\frac{1}{2l^4} \frac{d}{dl} \left[l^4 \frac{dS_2}{dl} \right] \quad (3.33)$$

Then we relate $\partial_k b_{k\parallel,\parallel}$ to $S_3(l)$. The incompressibility condition gives $0 = \partial_m' b_{ij,m}$. Using again Eqs. (3.16) and (3.18), we get

$$\left[\frac{dC}{dl} + \frac{2}{l}(C + D) \right] \delta_{ij} + \left[2 \frac{dD}{dl} + \frac{dF}{dl} + \frac{2}{l}(F - D) \right] n_i n_j = 0 \quad (3.34)$$

Thus

$$\frac{dC}{dl} + \frac{2}{l}(C + D) = 0 \quad (3.35)$$

$$\frac{d(3C + 2D + F)}{dl} + \frac{2}{l}(3C + 2D + F) = 0 \quad (3.36)$$

Equation (3.36) is obtained by taking the trace of Eq. (3.34). The functions C, D, and F have to be finite at $l = 0$, thus

$$3C + 2D + F = 0 \quad (3.37)$$

We can then express D and F in terms of C and dC/dl:

$$D = -C - \frac{l}{2}\frac{dC}{dl} \tag{3.38}$$

$$F = l\frac{dC}{dl} - C \tag{3.39}$$

and obtain

$$b_{ij,m} = C\delta_{ij}n_m - \left(C + \frac{l}{2}\frac{dC}{dl}\right)(\delta_{im}n_j + \delta_{jm}n_i) + \left(l\frac{dC}{dl} - C\right)n_in_jn_m \tag{3.40}$$

Thus

$$S_3(l) = \langle\delta u_\parallel^3\rangle = B_{ijm}n_in_jn_m \tag{3.41}$$

$$= 2(b_{ij,m} + b_{im,j} + b_{jm,i})n_in_jn_m = -12C(l) \tag{3.42}$$

where we have used $\langle U_i'U_j'U_m'\rangle = \langle U_iU_jU_m\rangle$. Moreover,

$$-\partial_k b_{ki,j} = \left(-\frac{2}{l}C + 2\frac{dC}{dl} + \frac{l}{2}\frac{d^2C}{dl^2}\right)n_in_j - \left(\frac{2}{l}C + 3\frac{dC}{dl} + \frac{l}{2}\frac{d^2C}{dl^2}\right)\delta_{ij}$$

$$\Rightarrow \quad -\partial_k b_{\parallel,\parallel} = -\frac{1}{l^4}\frac{d}{dl}(l^4C) = -\frac{1}{12l^4}\frac{d}{dl}(l^4S_3) \tag{3.43}$$

Putting all the results together, we finally obtain

$$\frac{1}{6l^4}\frac{d}{dl}(l^4S_3) - \frac{\nu}{l^4}\frac{d}{dl}\left(l^4\frac{dS_2}{dl}\right) = -\frac{2}{3}\langle\epsilon\rangle - \frac{1}{2}\partial_t S_2 + \langle\delta f_\parallel\delta u_\parallel\rangle \tag{3.44}$$

Here, we have used

$$\frac{1}{2}\partial_t\langle\vec{U}\cdot\vec{U}\rangle - \langle\vec{f}\cdot\vec{U}\rangle = -\langle\epsilon\rangle \tag{3.45}$$

which follows from Eq. (3.14). For decaying turbulence, $\vec{f} = 0$ and $\partial_t S_2 \approx 0$ for $l \ll l_0$. For stationary turbulence forced by $\vec{f}$ that acts only at the largest scales, $\partial_t S_2 = 0$ and $\langle\delta f_\parallel\delta u_\parallel\rangle \approx 0$ for $l \ll l_0$. Thus for both cases, we have

$$\frac{1}{6l^4}\frac{d}{dl}(l^4S_3) - \frac{\nu}{l^4}\frac{d}{dl}\left(l^4\frac{dS_2}{dl}\right) = -\frac{2}{3}\langle\epsilon\rangle \tag{3.46}$$

for $l \ll l_0$. In the limit of $\nu \to 0$, the viscous term is negligible. On the other hand, it is assumed that $\langle\epsilon\rangle$ remains finite in this limit. This implies that the velocity gradients $\partial u_i/\partial r_j$ become unlimited as $\nu \to 0$ or as Re $\to \infty$, which further implies that vorticity is generated in turbulent flows and increases with Re. The result that the dissipation remains finite as Re $\to \infty$ is generally referred to as the "dissipative anomaly", and is well supported by experimental and numerical results. Thus in the limit of $\nu \to 0$, integrating Eq. (3.46) gives the four-fifth law Eq. (3.7). We note that for stationary homogeneous and isotropic turbulence with $\vec{f}$ acting over all scales, we have the more general result [6]

$$S_3(l) = -\frac{4}{5}\langle\epsilon\rangle l + \frac{6}{l^4}\int_0^l l'^4 \langle\delta f_\parallel(l')\delta u_\parallel(l')\rangle dl' \qquad (3.47)$$

for $l \ll l_0$.

3.4 The Obukhov–Corrsin Theory for Passive Scalar

Obukhov [7] and Corrsin [8] extended Kolmogorov's 1941 theory to study temperature fluctuations in weakly-heated incompressible turbulent flows. The heating is so weak that the resulted temperature variations have no dynamical effect on the turbulent flow itself. As a result, the velocity field is still governed by the Navier–Stokes equation. In this case, the temperature is known as a passive scalar. The equations of motion are thus Eqs. (1.1) and (1.8). The Obukhov–Corrsin theory gives the statistics of the temperature difference, defined by

$$\delta T(\vec{r}, \vec{l}) \equiv T(\vec{r} + \vec{l}, t) - T(\vec{r}, t) \qquad (3.48)$$

which is taken to be statistically homogeneous and isotropic. Besides the cascade of turbulent energy, there is also a cascade of temperature variance from large to small scales. The mean temperature dissipation rate is given by $\langle\chi\rangle$, where $\chi(\vec{r}, t)$ is defined in Eq. (1.30). In analogy to the K41 theory, the temperature variance transfer rate, estimated by $[\delta T(l)]^2/t_l$, is scale-independent and thus equals to $\langle\chi\rangle$ in the intermediate inertial-convective range, the range of scales within the inertial range where buoyancy is insignificant. That is,

$$\frac{[\delta T(l)]^2 \delta u(l)}{l} \sim \langle\chi\rangle \qquad (3.49)$$

Together with Eq. (3.10) for $\delta u(l)$, we obtain

$$\delta T(l) \sim \langle\epsilon\rangle^{-1/6}\langle\chi\rangle^{1/2}l^{1/3} \qquad (3.50)$$

and the Obukhov–Corrsin (OC) scaling for passive temperature fluctuations:

$$S_p^\theta(l) \equiv \langle[\delta T(l)]^p\rangle \sim \langle\epsilon\rangle^{-p/6}\langle\chi\rangle^{p/2}l^{p/3} \qquad (3.51)$$

Here, S_p^θ is known as the pth order temperature structure functions. Experiments again confirm the power-law dependence but show that there are intermittency corrections to the OC scaling such that $S_p^\theta(l) \sim l^{\xi(p)}$ and $\xi(p)$ deviates from $p/3$ [9].

3.5 The Bolgiano–Obukhov Scaling

In turbulent convection, temperature variations result in a buoyancy force that drives the fluid motion, and temperature is now an active scalar. The presence of buoyancy could affect and modify the scaling behavior. In several theoretical studies [6, 10–13], arguments were given that buoyancy would give rise to a different scaling behavior:

$$S_p^u(l) \rangle \sim (\alpha g)^{2p/5} \langle \chi \rangle^{p/5} l^{3p/5} \tag{3.52}$$

$$S_p^\theta(l) \sim (\alpha g)^{-p/5} \langle \chi \rangle^{2p/5} l^{p/5} \tag{3.53}$$

This type of scaling behavior, which is known as the Bolgaino–Obukhov (BO) scaling, was originally proposed by Bolgiano [14] and Obukhov [15] for stably stratified flows (see also discussions in [16]) based on dimensional analysis and the argument that the velocity and temperature structure functions would depend only on αg, $\langle \chi \rangle$ and l. Here, αg is the additional parameter that describes the strength of buoyant coupling when buoyancy is significant. In turbulent Rayleigh–Bénard convection, the BO scaling can be obtained based on a cascade of temperature variance (Eq. (3.49)) or a cascade of entropy flux [12] (for $\theta \ll T_0$, $\int \theta^2 d^3 x$ describes the entropy increase per unit mass and volume due to the temperature fluctuations [13]) together with the argument that the buoyant term dominates the dynamics and balances the nonlinear advection term:

$$\alpha g \delta T(l) \sim \frac{[\delta u(l)]^2}{l} \tag{3.54}$$

Equations (3.49) and (3.54) imply

$$\delta u(l) \sim (\alpha g)^{2/5} \langle \chi \rangle^{1/5} l^{3/5} \tag{3.55}$$

$$\delta T(l) \sim (\alpha g)^{-1/5} \langle \chi \rangle^{2/5} l^{1/5} \tag{3.56}$$

Then Eqs. (3.52) and (3.53) follow directly.

3.6 Crossover in Scaling

The BO scaling would hold only when buoyancy is significant. When buoyancy is negligible, temperature behaves as a passive scalar and K41 and OC scaling would hold. The buoyant term, estimated by $\alpha g \delta T(l) \delta u(l)$, increases with l. Thus one expects a crossover from the K41-OC scaling to the BO scaling to occur at the crossover scale l_c when

$$\langle \delta u(l_c)^{BO} \rangle = \langle \delta u(l_c)^{K41} \rangle \tag{3.57}$$

Using Eqs. (3.10) and (3.55), we get

$$l_c = \frac{\langle \epsilon \rangle^{5/4}}{(\alpha g)^{3/2} \langle \chi \rangle^{3/4}} \equiv L_B \tag{3.58}$$

Thus the crossover scale is given by L_B, which is known as the Bolgiano length and is the length scale above which buoyancy is important. The Bolgiano length was first defined in terms of αg, $\langle \epsilon \rangle$, and $\langle \chi \rangle$ using dimensional analysis [16], Furthermore, we have

$$\alpha g \langle \delta u(l) \delta T(l) \rangle \geq \langle \epsilon \rangle \quad \text{for } l \geq L_B \tag{3.59}$$

therefore L_B is also the scale at which the power injected into the flow due to buoyancy is equal to the mean energy dissipation rate [17]. Using the exact relations Eqs. (1.41) and (1.42), L_B can be related to Nu and Ra:

$$L_B = \frac{\text{Nu}^{1/2}}{(\text{PrRa})^{1/4}} H \tag{3.60}$$

Hence, the picture emerging from these scaling theories is that the BO scaling is expected to hold in the buoyancy subrange, $l_0 \gg l > L_B$, while the K41-OC scaling is expected to hold in the inertial-convective subrange, $l_d \ll l < L_B$. If L_B is of the order of $l_0 \approx H$ or even larger, then only K41-OC scaling will be observed. On the other hand, if L_B is of the order of l_d or even smaller, then only the BO scaling would be observed [13]. However, there are two complications. The first complication is that turbulent Rayleigh–Bénard convection is inhomogeneous. Thus it is more appropriate to define a local crossover or Bolgiano length using the energy and thermal dissipation rates averaged over the local region of interest. As a result, it is possible that different scaling behavior is observed in different regions of the cell. This will be discussed in Chap. 4 when we examine the scaling behavior observed in experiments and numerical calculations. The second complication is the existence of intermittency corrections to the scaling behavior. In the next Section, we shall discuss one particular idea, the refined similarity hypothesis, which was proposed to account for the intermittency corrections.

3.7 Refined Similarity Hypothesis

To account for the intermittency corrections of velocity fluctuations, Kolmogorov proposed in 1962 [18] to refine his second hypothesis by replacing the mean energy dissipation rate $\langle \epsilon \rangle$ with a locally-averaged energy dissipation rate over a scale l, defined as

$$\epsilon_l(\vec{r}, t) \equiv \frac{3}{4\pi l^3} \int_{|\vec{y}| \le l} \epsilon(\vec{x} + \vec{y}, t) d\vec{y} \tag{3.61}$$

Similar ideas were also proposed independently by Obukhov [19]. As a result of this refinement, which is known as the refined similarity hypothesis (RSH), Eq. (3.10) is modified to

$$\delta u(l)^{K41} \sim \epsilon_l^{1/3} l^{1/3} \tag{3.62}$$

$$\Rightarrow \quad S_p^u(l) \sim \langle \epsilon_l^{p/3} \rangle l^{p/3} \tag{3.63}$$

Corrections to the K41 scaling can thus be resulted from the l-dependence of the moments of ϵ_l. In particular, let

$$\langle \epsilon_l^q \rangle \sim l^{\tau(q)} \tag{3.64}$$

then

$$\zeta(p) = \tau\left(\frac{p}{3}\right) + \frac{p}{3} \tag{3.65}$$

Different intermittency models have been proposed which give different results for $\tau(q)$.

A direct implication of Eq. (3.62) is

$$\langle [\delta u(l)]^p \mid \epsilon_l = x \rangle \sim x^{p/3} l^{p/3} \tag{3.66}$$

where $\langle [\delta u(l)]^p \mid \epsilon_l = x \rangle$ is the conditional velocity structure function of order p when the value of ϵ_l is fixed at a small range about x. Thus $\langle [\delta u(l)]^p \mid \epsilon_l = x \rangle \sim l^{p/3}$ exhibits the K41 scaling. Support for Eq. (3.66) has been found in both experiments [20] as well as in direct numerical simulations and large-eddy simulations [21].

The refined similarity hypothesis has been extended to temperature fluctuations by replacing also χ by the locally averaged $\chi_l(\vec{r}, t)$, which is similarly defined:

$$\chi_l(\vec{r}, t) = \frac{3}{4\pi l^3} \int_{|\vec{y}| \le l} \chi(\vec{x} + \vec{y}, t) d^3 y \tag{3.67}$$

For passive temperature fluctuations, Eq. (3.50) becomes [22, 23]:

$$\delta T(l)^{OC} \sim \epsilon_l^{-1/6} \chi_l^{1/2} l^{1/3} \tag{3.68}$$

and for the BO scaling, Eqs. (3.55) and (3.56) become [24]:

$$\delta u(l)^{BO} \sim (\alpha g)^{2/5} \chi_l^{1/5} l^{3/5} \tag{3.69}$$

$$\delta T(l)^{BO} \sim (\alpha g)^{-1/5} \chi_l^{2/5} l^{1/5} \tag{3.70}$$

3.8 Conditional Structure Functions

We note the interesting observation that the dependence on χ_l is different for the two scaling behaviors, K41-OC and BO, as shown in Eqs. (3.62), (3.68), (3.69) and (3.70). This difference can be clearly spelled out by studying the conditional velocity and temperature structure functions evaluated at fixed values of χ_l:

$$\tilde{S}_p^u(l, x) \equiv \langle [\delta u(l)]^p \mid \chi_l = x \rangle \tag{3.71}$$

$$\tilde{S}_p^\theta(l, x) \equiv \langle [\delta T(l)]^p \mid \chi_l = x \rangle \tag{3.72}$$

We have used these conditional structure functions [24] and similar conditional structure functions evaluated at given values of local temperature variance transfer rate [25] to examine the validity of refined similarity hypothesis in turbulent Rayleigh–Bénard convection.

To evaluate $\tilde{S}_p^u(l, x)$ and $\tilde{S}_p^\theta(l, x)$ from Eqs. (3.62) and (3.68) in the case of the K41-OC scaling, we need to evaluate the conditional average $\langle \epsilon_l^q \mid \chi_l = x \rangle$ for various values of q. In this case, temperature is a passive scalar so we make use of the measured approximate statistical independence of ϵ_l and χ_l for passive scalar fluctuations [26] to approximate:

$$\langle \epsilon_l^q \mid \chi_l = x \rangle \approx \langle \epsilon_l^q \rangle \qquad \text{K41} - \text{OC} \tag{3.73}$$

As a result, we obtain

$$\tilde{S}_p^u(l, x) \sim \begin{cases} \langle \epsilon_l^{p/3} \rangle l^{p/3} & \text{K41} \\ (\alpha g)^{2p/5} x^{p/5} l^{3p/5} & \text{BO} \end{cases} \tag{3.74}$$

$$\tilde{S}_p^\theta(l, x) \sim \begin{cases} \langle \epsilon_l^{-p/6} \rangle x^{p/2} l^{p/3} & \text{OC} \\ (\alpha g)^{-p/5} x^{2p/5} l^{p/5} & \text{BO} \end{cases} \tag{3.75}$$

From Eqs. (3.74) and (3.75), we see the different x-dependence of $\tilde{S}_p^u$ and $\tilde{S}_p^\theta$ for the two different scaling behaviors: $\tilde{S}_p^u$ is independent of x for the K41 scaling but has a power-law dependence of $x^{p/5}$ for the BO scaling. Similarly, $\tilde{S}_p^\theta$ has the power-law dependence of $x^{p/2}$ for the OC scaling but a different dependence of $x^{2p/5}$ for the BO scaling. Hence it is possible to reveal the two different scaling behaviors by studying the x-dependence of $\tilde{S}_p^u(l, x)$ and $\tilde{S}_p^\theta(l, x)$. This method is particularly useful because the unknown intermittency corrections might hinder direct revelation of the scaling behavior. Details about this method will be discussed in Chap. 4.

References

1. L.F. Richardson, *Weather Prediction by Numerical Process* (Cambridge University Press, Cambridge, 2007)
2. A.N. Kolmogorov, The local structure of turbulence in imcompressible viscous fluid for very large Reynolds numbers. C. R. (Dokl.) Acad. Sci. SSSR **30**, 301–305 (1941). Reprinted: (1991) Proc. R. Soc. Lond. Ser. A **434**, 9–13
3. A.N. Kolmogorov, Dissipation of energy in the locally isotropic turbulence. C. R. (Dokl.) Acad. Sci. SSSR 32:16–18 (1941). Reprinted: (1991) Proc. R. Soc. Lond. Ser. A **434**, 15–17
4. L.D. Landau, E.M. Lifshitz, *Fluid Mechanics* (Pergamon Press, Oxford, 1987)
5. U. Frisch, *Turbulence* (Cambridge University Press, Cambridge, 1995)
6. V. Yakhot, 4/5 Kolmogorov law for statistically stationary turbulence: application to High-Rayleigh-Number Bénard convection. Phys. Rev. Lett. **69**, 769–771 (1992)
7. A.M. Obukhov, The structure of the temperature field in a turbulent flow. Izv. Akad. Nauk. SSSR. Ser. Geogr. Geophys. **13**, 58–69 (1949)
8. S. Corrsin, On the spectrum of isotropic temperature fluctuations in isotropic turbulence. J. Appl. Phys. **22**, 469–473 (1951)
9. Z. Warhaft, Passive scalars in turbulent flows. Annu. Rev. Fluid Mech. **32**, 203–240 (2000)
10. I. Procaccia, R. Zeitak, Scaling exponents in nonisotropic convective turbulence. Phys. Rev. Lett. **62**, 2128–2131 (1989)
11. I. Procaccia, R. Zeitak, Scaling exponents in thermally driven turbulence. Phys. Rev. A **42**, 821–830 (1990)
12. V.S. L'vov, Spectra of velocity and temperature fluctuations with constant entropy flux of fully developed free-convective turbulence. Phys. Rev. Lett. **67**, 687–690 (1991)
13. S. Grossmann, V.S. L'vov, Crossover of spectral scaling in thermal turbulence. Phys. Rev. E **47**, 4161–4168 (1993)
14. R. Bolgiano, Turbulent spectra in a stably stratified atmosphere. J. Geophys. Res. **64**, 2226–2229 (1959)
15. A.M. Obukhov, The influence of Archimedean forces on the structure of the temperature field in a turbulent flow. Dokl. Akad. Nauk. SSR **125**, 1246–1248 (1959)
16. A.S. Monin, A.M. Yaglom, *Statistical Fluid Mechanics: Mechanics of Turbulence* (MIT Press, Cambridge, 1975)
17. E. Calzavarini, F. Toschi, R. Tripiccione, Evidences of Bolgiano-Obhukhov scaling in three-dimensional Rayleigh-Bénard convection. Phys. Rev. E **66**, 016304 (2002)
18. A.N. Kolmogorov, A refinement of previous hypotheses concerning the local structure of turbulence in a viscous incompressible fluid at high Reynolds number. J. Fluid Mech. **13**, 82–85 (1962)
19. A.M. Obukhov, J. Fluid Mech. **13**, 77 (1962)
20. A. Praskovsky, E. Praskovskaya, T. Horst, Further experimental support for the Kolmogorov refined similarity hypothesis. Phys. Fluids **9**, 2465–2467 (1997)
21. L.-P. Wang, S. Chen, J.G. Brasseur, J.C. Wyngaard, Examination of hypothesis in the Kolmogorov refined turbulence theory through high-resolution simulations. Part 1. Velocity field. J. Fluid Mech. **309**, 113–156 (1996)
22. G. Stolovitzky, P. Kailasnath, K.R. Sreenivasan, Refined similarity hypotheses for passive scalars mixed by turbulence. J. Fluid Mech. **297**, 275–291 (1995)
23. Y. Zhu, R.A. Antonia, I. Hosokawa, Refined similarity hypothesis for turbulent velocity and temperature fields. Phys. Fluids **7**, 1637–1648 (1995)
24. E.S.C. Ching, K.L. Chau, Conditional statistics of temperature fluctuations in turbulent convection. Phys. Rev. E **63**, 047303 (2001)
25. E.S.C. Ching, W.C. Cheng, Anomalous scaling and refined similarity of an active scalar in a shell model of homogeneous turbulent convection. Phys. Rev. E **77**, 015303(R) (2008)
26. G. Ruiz-Chavarria, C. Baudet, S. Ciliberto, Scaling laws and dissipation scale of a passive scalar in fully developed turbulence. Phys. D **99**, 369–380 (1996)

Chapter 4
Observed Scaling Behavior

Abstract We first introduce the local Bolgiano length, which depends on the vertical coordinate as a result of the inhomogeneity of the system. Based on the local Bolgiano length evaluated in numerical calculations, K41-OC scaling is expected in the central region and BO scaling is expected to exist only near the top and bottom plates. Then we discuss the experimentally observed scaling behavior in the central region, which has been reviewed in [1]. Next, we discuss the more recent analysis of the conditional temperature structure functions using experimental measurements at the bottom plate. We show that the experimental results are consistent with the theoretical expectations.

Keywords Local Bolgiano length · Conditional structure functions

4.1 Local Bolgiano Length

Turbulent Rayleigh–Bénard convection is not homogeneous. We make the assumption of statistical homogeneity and isotropy on a horizontal plane. Then one can define [2] a local Bolgiano length using $\langle \epsilon \rangle_z$ and $\langle \chi \rangle_z$:

$$l_B(z) = \frac{\langle \epsilon \rangle_z^{5/4}}{(\alpha g)^{3/2} \langle \chi \rangle_z^{3/4}} \tag{4.1}$$

Thus different scaling behavior could be observed in different regions of the convective cell depending on the relative size of $l_B(z)$ and H.

In two numerical studies using a lattice Boltzmann scheme [2, 3], $l_B(z)$ has been evaluated. Isothermal and no-slip boundary conditions are enforced respectively for the temperature and velocity fields at the top and bottom plates of the convection cell. The other boundary conditions are different in the two numerical studies. Adiabatic boundary conditions for the temperature field and free-slip boundary conditions for

E. S. C. Ching, *Statistics and Scaling in Turbulent Rayleigh-Bénard Convection*,
SpringerBriefs in Applied Sciences and Technology, DOI: 10.1007/978-981-4560-23-8_4,
© The Author(s) 2014

the velocity field were used on the vertical sidewalls in [2] whereas periodic boundary conditions were used in the horizontal directions (along the vertical sidewalls) for both fields in [3]. Despite the difference in the boundary conditions, the same general feature is found: $l_B(z)$ is maximum on the central plane $z = H/2$, and decreases as one moves toward the top and bottom plates: $z = H$ and $z = 0$. Specifically, for Ra $= 3.5 \times 10^7$ and Pr about 1, $l_B(z) > 0.8H$ in the central region $|z - H/2| \leq H/3$, and $l_B(z) < 0.2H$ for $z < 0.05H$ and $z > 0.95H$ [3].

In a more recent direct numerical simulation of an aspect-ratio-one cylindrical cell [4] for Ra ranges between 10^8 and 10^{10} with Pr $= 4$ or 6.4, adiabatic sidewall and isothermal top and bottom plates for the temperature field and no slip boundary condition for the velocity field at all boundaries were used. The pointwise Bolgiano length at each position $l_B(\vec{r})$ has been evaluated using the time-averaged energy and thermal dissipation rates:

$$l_B(\vec{r}) = \frac{\overline{\epsilon}(\vec{r})^{5/4}}{(\alpha g)^{3/2}\overline{\chi}(\vec{r})^{3/4}} \tag{4.2}$$

where $\overline{\epsilon}(\vec{r}) \equiv \langle \epsilon(\vec{r}, t) \rangle_t$ and $\overline{\chi}(\vec{r}) \equiv \langle \chi(\vec{r}, t) \rangle_t$. The numerical calculations show that $l_B(\vec{r})$ can be one order of magnitude larger than the global Bolgiano length L_B, calculated using Eq. (3.60) and the measured value of Nu, and is strongly space-dependent: $l_B(\vec{r})$ is the smallest (less than $0.1H$ at Ra $= 1 \times 10^9$) within the thermal boundary layers at the top and bottom plates, the average value of $l_B(\vec{r})$ in the central core region ($0.1H < r < 0.4H$ at $z = H/2$), denoted as l_B^{bulk}, is about $0.3H$ at Ra $\sim 10^{10}$, and $l_B(\vec{r})$ is the largest, comparable to H, at the mid-height of the sidewall. Moreover, l_B^{bulk} has been found to increase with Ra. The large $l_B(\vec{r})$ at the sidewall is not difficult to understand. Temperature gradient vanishes at the sidewall because of the thermally insulating boundary condition. As a result, $\overline{\chi}(\vec{r})$ is very small leading to the large value of $l_B(\vec{r})$ near the sidewall. On the other hand, the increase of l_B^{bulk} with Ra cannot be explained [1] using the estimates of the bulk quanities in the Grossmann-Lohse theory of heat transport [5]. Similarly, the relative size of $l_B(\vec{r})$ or $l_B(z)$ in the central bulk region and the thermal boundary layers is also remained to be understood.

Based on these numerical results of $l_B(z)$ and $l_B(\vec{r})$, one thus expects to observe the BO scaling only close to the top and bottom plates. Even in these boundary-layer regions, the BO scaling, if exists, would only cover a very narrow range because of the proximity of $l_B(z)$ to H. This lack of a long scaling range makes the direct observation of the BO scaling inherently challenging. Moreover, because the intermittency corrections to the scaling behavior are a priori unknown, it is additionally difficult to clearly identify the type of the scaling behavior observed. In the central region, to study scales larger than l_B^{bulk}, one would get close to the sidewall region at which $l_B(\vec{r})$ is the largest. Thus one expects to observe mostly the K41-OC scaling in the central region.

4.2 Early Measurements in Time Domain

In early experiments, measurements of the temperature fluctuations were taken by a thermistor and velocity fluctuations by laser Doppler velocimetry (LDV) as a function of time at a single point. Such measurements have been made at several positions within the convection cell. With the measured velocity and temperature time series, the temporal velocity and temperature differences between two times separated by a time interval τ:

$$\delta u(\tau) \equiv u(\vec{r}, t + \tau) - u(\vec{r}, t) \tag{4.3}$$

$$\delta T(\vec{r}, \tau) \equiv T(\vec{r}, t + \tau) - T(\vec{r}, t) \tag{4.4}$$

can be calculated. Then analogous to Eqs. (3.8) and (3.51), the temporal velocity and temperature structure functions can be defined as

$$S_p^{u,\tau}(\tau) \equiv \langle [\delta u(\tau)]^p \rangle \tag{4.5}$$

$$S_p^{\theta,\tau}(\tau) \equiv \langle [\delta T(\tau)]^p \rangle \tag{4.6}$$

using the temporal velocity and temperature differences. The frequency spectrum is the Fourier transform of the temporal correlation function (see Eq. (2.48)), thus

$$S_2^{u,\tau}(\tau) = 4 \int_0^\infty (1 - \cos f\tau) E_u(f) df \tag{4.7}$$

$$S_2^{\theta,\tau}(\tau) = 4 \int_0^\infty (1 - \cos f\tau) E_\theta(f) df \tag{4.8}$$

where $E_u(f)$ and $E_\theta(f)$ are the velocity and temperature frequency spectra. Thus if $S_2^{u,\tau} \sim \tau^\alpha$, $E_u(f) \sim f^{-(\alpha+1)}$ and vice versa. Similar result holds for $S_2^{\theta,\tau}(\tau)$ and $E_\theta(f)$. There were experimental reports [6–9] that $E_\theta(f) \sim f^{-7/5}$ at the cell center. The frequency spectrum of the vertical velocity component $E_u(f)$, measured at positions other than the cell center, at about $H/4$ from the bottom plate [7] and near the sidewall [10], was also found to exhibit a scaling in f: $E_u(f) \sim f^{-11/5}$. These observations thus imply $S_2^{u,\tau}(\tau) \sim \tau^{6/5}$ and $S_2^{\theta,\tau}(\tau) \sim \tau^{2/5}$. Taylor's frozen flow hypothesis [11] would relate $S_2^{u,\tau}(\tau)$ to $S_2^u(l)$, and $S_2^{\theta,\tau}(\tau)$ to $S_2^\theta(l)$ by $l = U_0 \tau$ where U_0 is the mean velocity of the turbulent flow. Thus the observed scaling in the frequency spectra apparently coincide with the BO scaling for $S_2^u(l)$ and $S_2^\theta(l)$ (see Eqs. (3.52) and (3.53)). However, Taylor's frozen flow hypothesis is valid only when the mean flow velocity is much larger than the root-mean-squared velocity fluctuation, and this condition is not satisfied in the central region. Moreover, using both velocity and temperature time-domain data at the cell center, Ching et al. [12] have found that the cross-scaling between the normalized temporal velocity and temperature structure functions is different from that implied by the BO scaling, casting further doubt to this apparent BO scaling at the cell center. Indeed, direct spatial measurements at the cell center showed that the scaling behavior at the cell center is not BO but K41-OC. This will be discussed in the next section. In particular, by comparing $S_3(l)$ and $S_3^{u,\tau}(\tau)$ directly, Sun et al. [13] have shown explicitly that the scaling exponent of $S_3^{u,\tau}(\tau)$ in τ is different from that of $S_3^u(l)$ in l.

4.3 Central Region

With developments in particle image velocimetry (PIV), direct measurements of the spatial velocity difference, between two points separated by a distance, have now been carried out. Using PIV and two thermistors with one fixed and the other movable along the vertical direction, Sun et al. [13] measured the velocity field in a vertical plane of 4×4 cm^2 in the central region, with the center of the plane located at the cell center, and the temperature difference along the vertical direction. The experiment was performed in a cylindrical cell of aspect ratio one with water. The velocity measurements were made at Ra $= 7.0 \times 10^9$ and temperature measurements at Ra $= 1.0 \times 10^{10}$ with Pr $= 4.3$ in both cases. Using the measured two-dimensional velocity field, they calculated the longitudinal velocity structure functions $S_p(l)$ as well as the transverse velocity structure functions, $S_p^\perp(l)$, with the separation perpendicular to the velocity direction. Their measurements show that both $S_p(l)$ and $S_p^\perp(l)$ obey the K41 scaling plus intermittency corrections close to that given by the She-Leveque model [14] while $S_p^\theta(l)$ obey OC scaling plus intermittency corrections close to those observed in a passive scalar experiment [15].

Later, Kunnen et al. [4] also measured spatial velocity structure functions using the stereoscopic PIV technique. This technique allows for measurements of the three velocity components at the same time. The experiment was done in water with Pr $= 6.37$ for three different values of Ra: 1.1×10^8, 3.34×10^8, and 1.10×10^9 and Kunnen et al. focused on $S_p^\perp(l)$ of the vertical velocity component. They estimated the value of l_B^{bulk} using the numerical data and found a very short range of the BO scaling at separations above l_B^{bulk} for the first and second-order transverse vertical velocity structure functions. For higher orders, the scaling exponents deviate from the BO values possibly due to intermittency corrections. They had also estimated $l_B^{bulk} \approx 5.6$ cm for Ra $= 1.0 \times 10^{10}$ and Pr $= 4.3$, explaining why Sun et al. did not observe the BO scaling in their experiment which covered only $l \leq 4$ cm. For separations below l_B^{bulk}, using extended self-similarity (ESS) [16], the relative scaling exponents are also consistent with the K41 scaling plus intermittency corrections given by the She-Leveque model.

Hence, direct spatial measurements show that the scaling behavior of velocity and temperature fluctuations in the central region is given by the K41-OC scaling, with the possibility of a very short range of the BO scaling for $l > l_B^{bulk}$ for certain range of Ra.

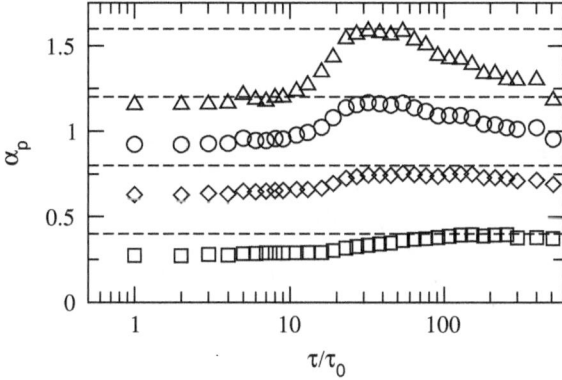

Fig. 4.1 Plot of α_p as a function of τ/τ_0 for measurements at the center of the bottom plate. $p = 1, 2, 3$ and 4 from bottom to top, and the *dashed lines* are $2p/5$

4.4 Near the Top and Bottom Plates

Kunnen et al. [4] have taken velocity measurements close to the top plate and found that the measurements were affected by the oscillating large-scale circulating flow. They did not observe BO scaling but found that the second-order transverse vertical velocity structure function has a scaling behavior consistent with the shear-flow scaling predicted by Lohse [17]:

$$S_p^u(l) \sim l^{2p/3} \tag{4.9}$$

$$S_p^\theta(l) \sim l^{p/6} \tag{4.10}$$

for a turbulent shear-dominated flow with passive temperature. Here s is the shear rate, which introduces an additional length scale and thus modifies Eqs. (3.8) and (3.51) to

$$\delta u(l) \sim \langle \epsilon \rangle^{1/6} s^{1/2} l^{2/3} \tag{4.11}$$

$$\delta T(l) \sim \langle \epsilon \rangle^{-1/12} s^{-1/4} \langle \chi \rangle^{1/2} l^{1/6} \tag{4.12}$$

On the other hand, Ching et al. [18] have attempted to reveal the scaling behavior using the conditional structure functions $\tilde{S}_p^u(l, x)$ and $\tilde{S}_p^\theta(l, x)$ at fixed local thermal dissipation rate $\chi_l = x$, as discussed in Sect. 3.8. To calculate these conditional velocity and temperature structure functions, one needs to have measurements of velocity, temperature and thermal dissipation rates taken simultaneously as a function of space and time. Such experimental measurements are very challenging and yet to be made. But simultaneous measurements of temperature and thermal dissipation rate at a fixed location as a function of time have been obtained recently [19]. The range of Ra covered is $9 \times 10^8 \leq \text{Ra} \leq 9 \times 10^9$ and $\text{Pr} = 5.5$. $\chi(\vec{r}, t)$ was measured

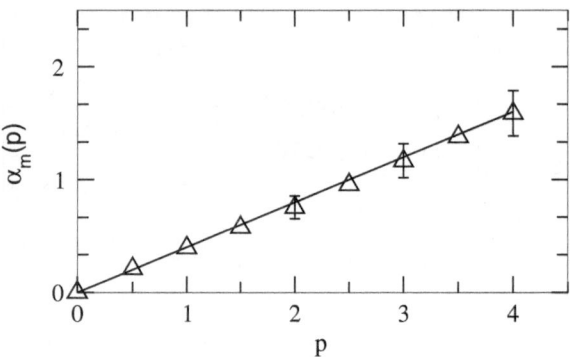

Fig. 4.2 Plot of $\alpha_m(p)$ as a function of p at the bottom plate (*triangles*). The *solid line* is $2p/5$

by a probe consisting of four small identical thermistors, with one placed at the center, and the other three placed at a distance $\delta l = 0.25 \pm 0.1$ mm from the central one, each along the three perpendicular directions. From the simultaneous temperature signals measured from the four thermistors, $T(t)$ from the central thermistor and $T_i(t) (i = x, y, z)$ from the other three thermistors, the three components of the temperature gradient, $(T_i - T)/\delta \ell$ ($i = x, y, z$), and thus $\chi(\vec{r}, t)$ can be obtained as a function of time t. Using these measurements, a locally averaged thermal dissipation rate over a time interval has been constructed [20, 21]:

$$\chi_\tau(\vec{r}, t) \equiv \frac{1}{\tau} \int\limits_t^{t+\tau} \chi_f(\vec{r}, t')dt \ , \tag{4.13}$$

where $\chi_f(\vec{r}, t) \equiv \kappa |\nabla T_f(\vec{r}, t)|^2$ and T_f is the temperature fluctuation. Then the conditional temperature structure functions in the time domain at a given fixed value of χ_τ:

$$\hat{S}_p^{\theta,\tau}(\tau, x) \equiv \langle |T(\vec{r}, t + \tau) - T(\vec{r}, t)|^p \mid \chi_\tau = x \rangle \tag{4.14}$$

are defined. The conditional structure functions, whether in spatial or time domain, have the same dimension, and χ_τ has the same dimension as χ_r, thus $\hat{S}_p^{\theta,\tau}(\tau, x)$ are expected to have the same power-law dependence on x as $\tilde{S}_p^\theta(l, x)$ (see Eq. (3.75)):

$$\hat{S}_p^{\theta,\tau}(\tau, x) \sim \begin{cases} x^{p/2} & \text{OC} \\ x^{2p/5} & \text{BO} \end{cases} \tag{4.15}$$

Since we are interested in the dependence on the value of χ_τ and not the scaling behavior, the difficulty of relating the scaling behavior in τ to that in l is avoided.

Using measurement taken at the center of the bottom plate, $\hat{S}_p^{\theta,\tau}(\tau, x)$ are calculated. In the calculation χ_τ is measured in units of the standard deviation σ_{χ_f} of

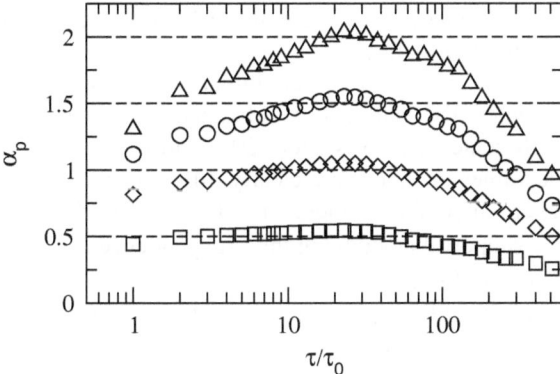

Fig. 4.3 Same as Fig. 4.1 for measurements at the cell center. The *dashed lines* are $p/2$

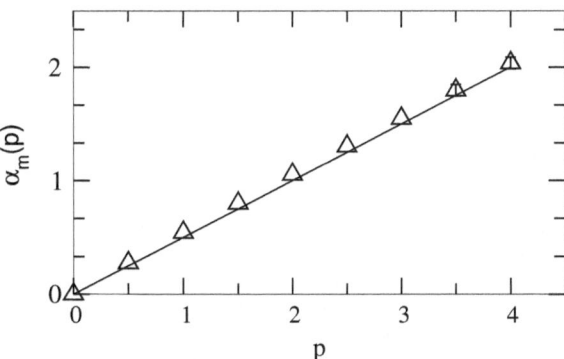

Fig. 4.4 Same as Fig. 4.2 for measurements at the cell center. The *solid line* is $p/2$

χ_f and the average is taken over those measurements with $|\chi_\tau/\sigma_{\chi_f} - x| \leq 0.005$. The value of τ is normalized by the sampling time interval $\tau_0 = 1/40\,s$. Power-law dependence on x has indeed been found:

$$\hat{S}_p^{\theta,\tau}(\tau, x) \sim x^{\alpha_p} \tag{4.16}$$

The values of α_p obtained are shown in Fig. 4.1. For each p, α_p attains a maximum value $\alpha_m(p)$ at a certain $\tau_m(p)$, therefore $\alpha_p \approx \alpha_m(p)$ for a small range of τ close to $\tau_m(p)$, i.e., we have

$$\hat{S}_p^{\theta,\tau}(\tau, x) \sim x^{\alpha_m(p)} \quad \text{for } \tau \approx \tau_m(p) \tag{4.17}$$

It is found that $\tau_m(p)$ decreases from about $200\tau_0$ to about $20\tau_0$ as p increases, indicating that the range of τ over which Eq. (4.17) holds changes with p. This feature has not been understood. We plot $\alpha_m(p)$ as a function of p in Fig. 4.2.

It can be seen that the values of $\alpha_m(p)$ are in excellent agreement with the predicted values of $2p/5$ for the BO scaling.

As a comparison, we have carried out similar analysis using the measurements taken at the cell center. The values of α_p obtained at the cell center are shown in Fig. 4.3. At the cell center, $\tau_m(p)$ is about $20\tau_0$ and is the same for all the values of p studied. The values of $\alpha_m(p)$ are clearly different from those obtained at the bottom plate, and are in good agreement with $p/2$ (see Fig. 4.4) for the OC scaling. These results thus support that studying the x-dependence of $\hat{S}_p^{\theta,\tau}(\tau, x)$ can reveal the scaling behavior of $S_p^\theta(l)$.

In short, the analysis of the χ_τ-dependence of the conditional temperature structure functions $\hat{S}_p^{\theta,\tau}(\tau, x)$ has successfully recovered the OC scaling plus intermittency corrections observed experimentally in the central region. Moreover, the observed χ_τ-dependence at the bottom plate indicates that the scaling behavior near the bottom plate is likely to be BO scaling plus intermittency corrections.

References

1. D. Lohse, K.-Q. Xia, Small-scale properties of turbulent Rayleigh-Béard convection. Annu. Rev. Fluid Mech. **42**, 335–364 (2010)
2. R. Benzi, F. Toschi, R. Tripiccione, On the heat transfer in Rayleigh-Bénard systems. J. Stat. Phys. **93**, 901–918 (1998)
3. E. Calzavarini, F. Toschi, R. Tripiccione, Evidences of Bolgiano-Obhukhov scaling in three-dimensional Rayleigh-Bénard convection. Phys. Rev. E **66**, 016304 (2002)
4. R.J.P. Kunnen, H.J.H. Clercx, B.J. Geurts et al., A numerical and experimental investigation of structure function scaling in turbulent Rayleigh-Bénard convection. Phys. Rev. E **77**, 016302 (2008)
5. S. Grossmann, D. Lohse, Scaling in thermal convection: a unifying view. J. Fluid. Mech. **407**, 27–56 (2000)
6. X.-Z. Wu, L. Kadanoff, A. Libchaber, M. Sano, Frequency power spectrum of temperature fluctuations in free convection. Phys. Rev. Lett. **64**, 2140–2143 (1990)
7. S. Ashkenazi, V. Steinberg, Spectra and statistics of velocity and temperature fluctuations in turbulent convection. Phys. Rev. Lett. **83**, 4760–4763 (1999)
8. J.J. Niemiela, L. Skrbek, K.R. Sreenivasan, R.J. Donnelly, Turbulent convection at very high Rayleigh numbers. Nature (London) **404**, 837–840 (2000)
9. S.-Q. Zhou, K.-Q. Xia, Scaling properties of the temperature field in convective turbulence. Phys. Rev. Lett. **87**, 064501 (2001)
10. X.-D. Shang , K.-Q. Xia, Scaling of the velocity power spectra in turbulent thermal convection. Phys. Rev. E **64**, 065301(R) (2001)
11. G.I. Taylor, The spectrum of turbulence. Proc. R. Soc. Lond. A **164**, 476–490 (1938)
12. E.S.C. Ching, K.W. Chui, X.-D. Shang et al., Velocity and temperature cross-scaling in turbulent thermal convection. J. Turbul. **5**, 27 (2004)
13. C. Sun, Q. Zhou, K.-Q. Xia, Cascades of velocity and temperature fluctuations in buoyancy-driven thermal turbulence. Phys. Rev. Lett. **97**, 144504 (2006)
14. Z.-S. She, E. Leveque, Universal scaling laws in fully developed turbulence. Phys. Rev. Lett. **72**, 336–339 (1994)
15. G. Ruiz-Chavarria, C. Baudeta, S. Ciliberto, Scaling laws and dissipation scale of a passive scalar in fully developed turbulence. Phys. D **99**, 369–380 (1996)

16. R. Benzi, S. R. Ciliberto, R. Tripiccione et al., Extended self-similarity in turbulent flows. Phys. Rev. E **48**, R29–R32 (1993)
17. D. Lohse, Temperature spectra in shear flow and thermal convection. Phys. Lett. A **196**, 70–75 (1994)
18. E.S.C. Ching, Y.K. Tsang, T.N. Fok et al., Scaling behavior in turbulent Rayleigh-Bénard convection revealed by conditional structure functions. Phys. Rev. E **87**, 013005 (2013)
19. X. He, P. Tong, Measurements of the thermal dissipation field in turbulent Rayleigh-Bénard convection. Phys. Rev. E **79**, 026306 (2009)
20. X. He, P. Tong, E.S.C. Ching, Statistics of the locally-averaged thermal dissipation rate in turbulent Rayleigh-Bénard convection. J. Turbul. **11**, 1 (2010)
21. X. He, E.S.C. Ching, P. Tong, Locally averaged thermal dissipation rate in turbulent thermal convection: a decomposition into contributions from different temperature gradient components. Phys. Fluids **23**, 025106 (2011)

Chapter 5
Summary and Outlook

We summarize our present understanding of the statistical properties and scaling behavior of turbulent Rayleigh-Bénard convection, and discuss the outstanding issues to be understood in future work.

The Rayleigh-Bénard convection system consists of a closed cell of fluid heated from below and cooled from above. Turbulent Rayleigh-Bénard convection is a fundamental problem of great research interest. In this monograph, we have discussed Rayleigh-Bénard convection exclusively in the Oberbeck-Boussinesq approximation, and focused on two issues of interest. The first issue is the characterization and understanding of the statistics of the velocity and temperature fluctuations in the system. The second issue is the revelation and understanding of the nature of the scaling behavior of the velocity and temperature structure functions.

The statistics of the velocity and temperature fluctuations are characterized by their PDFs. Because of the closure problem in turbulence, these PDFs cannot be calculated directly from the equations of motion. When the fluctuations obey certain statistical symmetries such as stationarity or statistical homogeneity, there are exact implicit PDF formulae that express the PDF of a fluctuation in terms of two conditional means of its time derivative or spatial gradients. Although these exact formulae cannot give us explicit results for the PDF, they could provide insights for understanding the features of the PDF of the fluctuations. In turbulent Rayleigh-Bénard convection, velocity and temperature fluctuations are stationary but generally statistically inhomogeneous. In the central region, the PDF of the normalized temperature fluctuation $X(t) = [T(t) - \langle T \rangle]/\langle [T(t) - \langle T \rangle]^2 \rangle^{1/2}$, denoted by $P_X(x)$, is Gaussian at low Ra but displays exponential tails at higher Ra. In addition, there is the interesting result that the conditional mean $\langle \ddot{X} \mid X = x \rangle \approx -x$. Using this result together with the exact PDF formula for stationary fluctuations [Eq. (2.76)], the change in the functional form of $P_X(x)$ can be understood as an increase in the statistical correlation of $\dot{X}(t)$ with $X(t)$ as Ra increases, which is manifested as the change of the conditional mean $\langle (\dot{X})^2 \mid X = x \rangle$ from being x-independent to having a linear dependence on $|x|$. Within the thermal boundary layers near the top and bottom plates of the convective cell, $P_X(x)$ becomes highly asymmetric.

E. S. C. Ching, *Statistics and Scaling in Turbulent Rayleigh-Bénard Convection*,   61
SpringerBriefs in Applied Sciences and Technology, DOI: 10.1007/978-981-4560-23-8_5,
© The Author(s) 2014

This skewness naturally occurs as a consequence of the much more frequent hotter or colder fluctuations from the bottom or the top plate. The feature $\langle \ddot{X} \mid X = x \rangle \approx -x$ does not hold and P_X depends on both $\langle \ddot{X} \mid X = x \rangle$ and $\langle (\dot{X})^2 \mid X = x \rangle$. Making the assumption that the fluctuations on a horizontal plane are statistically homogeneous and isotropic and using the equation of motion for the temperature field [Eq. (1.8)], $P_\theta(w)$ can be related to the conditional mean of the vertical velocity $\langle U_z \mid \theta = w \rangle_z$ and the conditional mean of the thermal dissipation $\kappa \langle \nabla^2 T \mid \theta = w \rangle_z$, evaluated at a given temperature fluctuation. Both of these two conditional means are related to the dynamical features of the flow. In particular, the conditional mean $\langle U_z \mid \theta = w \rangle_z$ has been naturally identified as the vertical velocity of the thermal plumes [1]. Thus the thermal plumes are a source of the skewness of the temperature fluctuations within the thermal boundary layer. More generally, since $X(t)$ and $\theta(t)$ are related, the conditional means $\langle \ddot{X} \mid X = x \rangle$ and $\langle (\dot{X})^2 \mid X = x \rangle$ are related to the conditional means $\langle U_z \mid \theta = w \rangle_z$ and $\kappa \langle \nabla^2 T \mid \theta = w \rangle_z$ with $w = x \langle [T(t) - \langle T \rangle_z]^2 \rangle_z^{1/2} + \langle T \rangle_z - T_0$. In future work, it would be interesting to study these relations and to understand the statistical properties in terms of the dynamical features of the flow. In the early days, velocity measurements were taken using LDV, and LDV measurements are not sampled uniformly in time, making it difficult to obtain velocity time derivatives directly from the measurements. As a result, previous studies have concentrated on temperature fluctuations and the conditional means of the temperature derivatives. With the advance in PIV measurements and in direct numerical simulations, extension of the study to velocity fluctuations and the conditional means of the velocity derivatives is possible and could lead to additional insights.

For the scaling behavior of velocity and temperature structure functions, the picture emerging from the scaling theories is that: the BO scaling plus intermittency corrections holds for scales above the Bolgiano length while the K41-OC scaling plus intermittency corrections holds for scales below the Bolgiano length. The relevant Bolgiano length, which is the crossover scale of the scaling behavior, is the local Bolgiano length evaluated using the local mean energy and thermal dissipation rates. Numerical simulations show that the local Bolgiano length $l_B(z)$ is highly inhomogeneous in turbulent Rayleigh-Bénard convection: it is maximum and of the order of the integral scale H on the central plane and is minimum close to the top and bottom plates. In the central region, experimental measurements [2] show that the velocity structure functions, both longitudinal and transverse, obey the K41 scaling plus intermittency corrections close to that given by the She-Leveque model [3] while the temperature structure functions obey OC scaling plus intermittency corrections close to those observed in a passive scalar experiment [4]. This observed K41-OC scaling behavior is thus consistent with the relatively large $l_B(z)$ in the central region. More recent measurements [5] showed that a very short range of the BO scaling is possible for $l > l_B(z)$ for some Ra. On the other hand, by studying the dependence of the conditional temperature structure functions on the locally averaged thermal dissipation rate, indication of the BO scaling plus intermittency corrections has indeed been found at the bottom plate [6]. Thus the observed scaling behavior of Rayleigh-Bénard convection is in accord with the theoretical picture. In future work, it would be interesting to directly observe the BO scaling for both velocity and temperature fluctuations. For this purpose, measurements taken on a horizontal plane as close to the top and bottom plates as possible are needed. Such velocity measurements are difficult to obtain experimentally. Investigation using DNS calculations with the boundary condition of constant heat flux [Eq. (1.13)] would thus be fruitful.

We end this Chapter by discussing two questions that remain to be answered. The first question is what physics sets the size of the local Bolgiano length. Specifically, why is the local Bolgiano length in turbulent Rayleigh-Bénard convection comparable to H for most part of the convective cell and is relatively smaller than H only near the top and bottom boundaries? Insights can be gained by looking at the exact relations discussed in Sect. 1.4. The exact balance Eq. (1.31) reads:

$$\alpha g \langle v_z \theta \rangle_{V,t} = \alpha g \langle u_z \theta \rangle_{V,t} = \langle \epsilon \rangle_{V,t} \tag{5.1}$$

For homogeneous systems, one might take $\alpha g \langle u_z \theta \rangle_{V,t}$ loosely as the sum of $\alpha g \langle \delta u(l) \delta \theta(l) \rangle_t$ over all the scales $l_d \leq l \leq l_0 = H$, and get

$$\sum_{l=l_d}^{H} \alpha g \langle \delta u(l) \delta \theta(l) \rangle_t \sim \langle \epsilon \rangle_{V,t} \tag{5.2}$$

One expects the correlation $\langle \delta u(l) \delta \theta(l) \rangle_t$ to increase with l and be positive in thermal convection such that the sum is dominated by the contribution at the integral scale $l_0 \sim H$, which implies

$$\alpha g \langle \delta u(H) \delta \theta(H) \rangle_t \sim \langle \epsilon \rangle_{V,t} \tag{5.3}$$

giving $L_B \approx H$ using Eq. (3.59). Homogeneous turbulent Rayleigh-Bénard convection [7] can be obtained numerically by imposing periodic boundary conditions for both the velocity and temperature at the top and bottom boundaries, and driving the system by a constant temperature gradient along the vertical direction. In three-dimensional homogeneous turbulent Rayleigh-Bénard convection, K41-OC scaling plus intermittency corrections has indeed been observed [8]. A completely analogous relation for Eq. (5.2) can be derived [9] for the shell model of homogeneous turbulent convection. In two dimensions, because of the inverse energy transfer from small to large scales, a damping has to be enforced at the integral scale to achieve stationarity. In this case, the exact balance would be modified to:

$$\alpha g \langle u_z \theta \rangle_{V,t} = \langle \epsilon \rangle_{V,t} + \langle \epsilon_{damp} \rangle_{V,t} \tag{5.4}$$

where $\langle \epsilon_{damp} \rangle_{V,t}$ is the mean energy dissipation rate due to the damping. When $\langle \epsilon_{damp} \rangle_{V,t} \gg \langle \epsilon \rangle_{V,t}$, then we can have $L_B \ll H$ and thus the BO scaling. This has indeed been found in two-dimensional homogeneous turbulent Rayleigh-Bénard convection, in which the BO scaling plus intermittency corrections has been observed [10]. The BO scaling has also been observed in two-dimensional experiments using soap films [11] and soap bubbles [12, 13]. In future work, it would be interesting to make these heuristic results for the local Bolgiano length rigorous for turbulent Rayleigh-Bénard convection confined by physical boundary conditions.

The second question is whether the BO scaling is associated with an upscale energy transfer [14]. In the BO scaling, Eq. (3.55) implies that the rate of energy transfer is scale-dependent:

$$\frac{[\delta u(l)]^3}{l} \sim l^{4/5} \tag{5.5}$$

This is feasible if there is a conversion between buoyant potential energy and turbulent kinetic energy at all scales at which the BO scaling holds. This picture is in contrast to the Richardson energy-cascade picture in which energy is input at the largest scale only and the mean rate of energy transfer is constant in the inertial range. In Rayleigh-Bénard convection, the thermal stratification is unstable and we, therefore, expect a net conversion of buoyant potential energy into turbulent kinetic energy due to work done by the buoyant forces. Eq. (5.5) indicates that the rate of energy transfer decreases as the scale decreases. If energy is transferred from large to small scales, it is difficult to reconcile the decreasing rate of energy transfer as one goes from large to small scales with the injection of turbulent energy from buoyant forces at all scales. This difficulty, referred to as the "paradoxical nature" of the BO scaling, was discussed as an argument against the realization of the BO scaling [15]. Note that such a difficulty does not exist for turbulent flows with stable thermal stratification since in this case, there is a net conversion of turbulent energy into buoyant potential energy through work done by the turbulent motion against the buoyant forces. This is why the BO scaling was first proposed for stably stratified flows [16]. Here, we point out the interesting possibility that this apparent difficulty in Rayleigh-Bénard convection can be avoided if energy is actually transferred from small to large scales when the BO scaling is valid. When turbulent energy is transferred from small to large scales, the injection of kinetic energy into the flow due to the work done by the buoyant forces against the turbulent motion would consistently give rise to an energy transfer rate that is increasing with the scale as depicted by Eq. (5.5). Hence, it would be interesting to investigate the direction of energy transfer to check the validity of this picture in future work.

References

1. E.S.C. Ching, H. Guo, X.D. Shang et al., Extraction of plumes in turbulent thermal convection. Phys. Rev. Lett. **93**, 124501 (2004)
2. C. Sun, Q. Zhou, K.-Q. Xia, Cascades of velocity and temperature fluctuations in buoyancy-driven thermal turbulence. Phys. Rev. Lett. **97**, 144504 (2006)
3. Z.-S. She, E. Leveque, Universal scaling laws in fully developed turbulence. Phys. Rev. Lett. **72**, 336–339 (1994)
4. G. Ruiz-Chavarria, C. Baudeta, S. Ciliberto, Scaling laws and dissipation scale of a passive scalar in fully developed turbulence. Physica. D **99**, 369–380 (1996)

5. R.J.P. Kunnen, H.J.H. Clercx, B.J. Geurts et al., A numerical and experimental investigation of structure function scaling in turbulent Rayleigh-Bénard convection. Phys. Rev. E **77**, 016302 (2008)
6. E.S.C. Ching, Y.K. Tsang, T.N. Fok et al., Scaling behavior in turbulent Rayleigh-Bénard convection revealed by conditional structure functions. Phys. Rev. E **87**, 013005 (2013)
7. V. Borue, S.A. Orszag, Turbulent convection driven by a constant temperature gradient. J. Sci. Comput. **12**, 305–351 (1995)
8. L. Biferale, E. Calzavarini, F. Toschi, R. Tripiccione, Universality of anisotropic fluctuations from numerical simulations of turbulent flows. Europhys. Lett. **64**, 461–467 (2003)
9. E.S.C. Ching, T.C. Ko, Ultimate-state scaling in a shell model for homogeneous turbulent convection. Phys. Rev. E **78**, 036309 (2008)
10. A. Celani, T. Matsumoto, A. Mazzino, M. Vergassola, Scaling and universality in turbulent convection. Phys. Rev. Lett. **88**, 054503 (2002)
11. J. Zhang, X.L. Wu, Velocity Intermittency in a buoyancy subrange in a two-dimensional soap film convection experiment. Phys. Rev. Lett. **94**, 234501 (2005)
12. F. Seychelles, Y. Amarouchene, M. Bessafi, H. Kellay, Thermal convection and emergence of isolated vortices in soap bubbles. Phys. Rev. Lett. **100**, 144501 (2008)
13. F. Seychelles, F. Ingremeau, C. Pradere, H. Kellay, From intermittent to nonintermittent behavior in two dimensional thermal convection in a Soap Bubble. Phys. Rev. Lett. **105**, 264502 (2010)
14. G. Boffetta, F. De Lillo, A. Mazzino, S. Musacchio, Bolgiano scale in confined Rayleigh-Taylor turbulence. J. Fluid Mech. **690**, 426–440 (2012)
15. D. Lohse, K.-Q. Xia, Small-scale properties of turbulent Rayleigh-Béard convection. Annu. Rev. Fluid Mech. **42**, 335–364 (2010)
16. A.S. Monin, A.M. Yaglom, *Statistical Fluid Mechanics: Mechanics of Turbulence* (MIT Press, Cambridge, Massachusetts, 1975)